Selected Titles in This Series

275 **Alfred G. Noël, Earl Barnes, and Sonya A. F. Stephens, Editors,** Council for African American researchers in the mathematical sciences: Volume III, 2001

274 **Ken-ichi Maruyama and John W. Rutter, Editors,** Groups of homotopy self-equivalences and related topics, 2001

273 **A. V. Kelarev, R. Göbel, K. M. Rangaswamy, P. Schultz, and ... er, Editors,** Abelian groups, rings and modules, 2001

272 **Eva Bayer-Fluckiger, David Lewis, and Andrew ...** forms and their applications, 2000

271 **J. P. C. Greenlees, Robert R. Bruner, and ...** methods in algebraic topology, 2001

270 **Jan Denef, Leonard Lipschitz, Thanases Pheid... ...el, Editors,** Hilbert's tenth problem: Relations with arithmetic and ... etry, 2000

269 **Mikhail Lyubich, John W. Milnor, and Yair N. ... ky, Editors,** Laminations and foliations in dynamics, geometry and topology, 2001

268 **Robert Gulliver, Walter Littman, and Roberto Triggiani, Editors,** Differential geometric methods in the control of partial differential equations, 2000

267 **Nicolás Andruskiewitsch, Walter Ricardo Ferrer Santos, and Hans-Jürgen Schneider, Editors,** New trends in Hopf algebra theory, 2000

266 **Caroline Grant Melles and Ruth I. Michler, Editors,** Singularities in algebraic and analytic geometry, 2000

265 **Dominique Arlettaz and Kathryn Hess, Editors,** Une dégustation topologique: Homotopy theory in the Swiss Alps, 2000

264 **Kai Yuen Chan, Alexander A. Mikhalev, Man-Keung Siu, Jie-Tai Yu, and Efim I. Zelmanov, Editors,** Combinatorial and computational algebra, 2000

263 **Yan Guo, Editor,** Nonlinear wave equations, 2000

262 **Paul Igodt, Herbert Abels, Yves Félix, and Fritz Grunewald, Editors,** Crystallographic groups and their generalizations, 2000

261 **Gregory Budzban, Philip Feinsilver, and Arun Mukherjea, Editors,** Probability on algebraic structures, 2000

260 **Salvador Pérez-Esteva and Carlos Villegas-Blas, Editors,** First summer school in analysis and mathematical physics: Quantization, the Segal-Bargmann transform and semiclassical analysis, 2000

259 **D. V. Huynh, S. K. Jain, and S. R. López-Permouth, Editors,** Algebra and its applications, 2000

258 **Karsten Grove, Ib Henning Madsen, and Erik Kjær Pedersen, Editors,** Geometry and topology: Aarhus, 2000

257 **Peter A. Cholak, Steffen Lempp, Manuel Lerman, and Richard A. Shore, Editors,** Computability theory and its applications: Current trends and open problems, 2000

256 **Irwin Kra and Bernard Maskit, Editors,** In the tradition of Ahlfors and Bers: Proceedings of the first Ahlfors-Bers colloquium, 2000

255 **Jerry Bona, Katarzyna Saxton, and Ralph Saxton, Editors,** Nonlinear PDE's, dynamics and continuum physics, 2000

254 **Mourad E. H. Ismail and Dennis W. Stanton, Editors,** q-series from a contemporary perspective, 2000

253 **Charles N. Delzell and James J. Madden, Editors,** Real algebraic geometry and ordered structures, 2000

252 **Nathaniel Dean, Cassandra M. McZeal, and Pamela J. Williams, Editors,** African Americans in Mathematics II, 1999

(Continued in the back of this publication)

Row 1 (Front, left to right)**:** 1. Monica Jackson; 2. Brian Rider; 3. Herbert Winful; 4. William A. Massey; 5. Frank Ingram; 6. J. Ernest Wilkins Jr.; 7. Kymberly Riggins; 8. Demthria Johnson; 9. Dennis Dean; 10. Kang Su Gatlin; 11. Jakita N. Owensby; 12. Juan E. Gilbert; 13. Olu Ogunsanya; **Row 2:** 14. Robert Megginson; 15. Arthur Grainger; 16. Kathy Lewis; 17. Gaston N'Guérékata; 18. Janis M. Oldham; 19. Annette Aaron; 20. Jacqueline Moore; 21. Rudy Horne; 22. Julie Ivy; 23. Lee Lorch; 24. Jeffrey Forbes; 25. Jean Cadet; 26. Roscoe Giles; **Row 3:** 27. Byron Jeff; 28. Judy Cassamajor; 29. William Christian; 30. Johnny E. Brown; 31. Courtney Davis; 32. Monica Stephens; 33. Janice Walker; 34. Johnny L. Houston; 35. Shaun Gittens; 36. Donald King; 37. Bryon Grayson; 38. Samuel Jator; 39. Mark Lewis; **Row 4:** Asamoah Nkwanta; 41. Angela Grant; 42. John Perkins; 43. Ahmed Ridley; 44. Semmi Pasha; 45. Carl Cowen; 46. Charles Hardnett; 47. Melvin Currie; 48. Damon Tull; 49. Jack Alexander; 50. Madison Gray; 51. Aris Winger; 52. Otis Jennings; 53. Charles Isbell; **Row 5:** 54. Carl Graham; 55. James Donaldson; 56. William Hawkins; 57. Desmond Stephens; 58. Scott Williams; 59. Illya Hicks; 60. Leon Woodson; 61. Rodrigo Bañuelos; 62. Jonathan Farley; 63. Alfred Noël.

CONTEMPORARY MATHEMATICS

275

Council for African American Researchers in the Mathematical Sciences: Volume III

CAARMS

Fifth Conference for African American Researchers in the Mathematical Sciences
June 22–25, 1999
University of Michigan, Ann Arbor, Michigan

Alfred G. Noël
Earl Barnes
Sonya A. F. Stephens
Editors

American Mathematical Society
Providence, Rhode Island

This volume contains research and expository papers presented primarily at the CAARMS5 (5th Conference for African American Researchers in the Mathematical Sciences) meeting held June 22–25, 1999 at the University of Michigan in Ann Arbor, Michigan and supplemented by presentations made at the CAARMS3 meeting held June 17–20, 1997 at Morgan State University in Baltimore, Maryland. Financial support was provided by the National Security Agency, the University of Michigan, and Morgan State University.

2000 *Mathematics Subject Classification.* Primary 33F05, 47N40, 68R05, 68R10, 65F25, 11R37, 60K20, 60K25, 68M20, 00A30.

Library of Congress Cataloging-in-Publication Data

Conference for African American Researchers in the Mathematical Sciences (5th : 1999 : Ann Arbor, Mich.)

Council for African American Researchers in the Mathematical Sciences. Volume III: Fifth Conference for African American Researchers in the Mathematical Sciences, June 22–25, 1999, University of Michigan, Ann Arbor, Michigan / Alfred G. Noël, Earl Barnes, Sonya A. F. Stephens, editors.

p. cm. — (Contemporary mathematics, ISSN 0271-4132 ; 275)

Includes bibliographical references.

ISBN 0-8218-2141-5 (alk. paper)

1. Mathematics—Congresses. I. Noël, Alfred G., 1956– II. Barnes, Earl. III. Stephens, Sonya A. F., 1970– IV. Council for African American Researchers in the Mathematical Sciences. V. Conference for African American Researchers in the Mathematical Sciences (3rd : 1997 : Baltimore, Md.) VI. Title. VII. Contemporary mathematics (American Mathematical Society) ; v. 275.

QA1 .C6253 1999
510—dc21 2001022099

Printed in the United States of America.

∞ The paper used in this book is acid-free and falls within the guidelines
established to ensure permanence and durability.

10 9 8 7 6 5 4 3 2 1   06 05 04 03 02 01

Contents

Preface

The Council for African American Researchers in the Mathematical Sciences (CAARMS) is a group dedicated to organizing an annual conference that showcases the current research primarily, but not exclusively, of African Americans in the mathematical sciences, which includes mathematics, operations research, statistics, and computer science. Held annually since 1995, the Conference for African American Researchers in the Mathematical Sciences (also called CAARMS) has had significant numbers of African American researchers presenting their work in hour long technical presentations and African American graduate students presenting their current work in an organized poster session. These events create a forum for mentoring and networking, where attendees can meet an audience of African American researchers and graduate students interested in such fields. The members of the council work on other issues such as strategies for increasing the number of doctoral recipients from underrepresented minority groups.

Although the current proceedings are considered to be "Volume III" of this series, Volumes I and II were published under different titles. Volume I is called "African Americans in Mathematics" and contains proceedings from the second CAARMS meeting, called CAARMS2, held June 25–28, 1996 at the Center for Discrete Mathematics and Computer Science (DIMACS) in Piscataway, New Jersey. It is also Volume 34 of the Series in Discrete Mathematics and Theoretical Computer Science. Volume II is called "African Americans in Mathematics II" and contains the proceedings from the CAARMS4 meeting, held June 16–19, 1998 at Rice University, in Houston, Texas. It is Volume 252 of the Contemporary Mathematics Series of the American Mathematical Society. Henceforth, the current title of Volume III, "Council for African American Researchers in the Mathematical Sciences" will be the permanent title for the series. This volume includes research and expository papers presented at both the CAARMS3 meeting held June 17–20, 1997 at Morgan State University in Baltimore, Maryland and at the CAARMS5 meeting held June 22–25 1999 at the University of Michigan in Ann Arbor, Michigan. The work covers a wide range of mathematics and every reader will find something of interest. Some papers are comprehensive surveys of research topics for which no text book exists. Other papers develop mathematics to obtain a deeper theoretical insight into mathematics or to solve problems in manufacturing and oceanography. Finally, some papers discuss new ways to view the totality of mathematics and move beyond the "pure" versus "applied" paradigm. The general style is rather expository which should help the reader.

Those meriting special thanks include the organizers of the CAARMS5 meeting: William Massey from Bell Laboratories of Lucent Technologies, Leon Woodson from Morgan State University, and our University of Michigan hosts John Birge, currently the Dean of Engineering at Northwestern University, and Robert Megginson of the University of Michigan. William Massey is one of the co-founders of CAARMS and has been one of the main organizers of every CAARMS meeting. John Birge and Robert Megginson respectively represent the joint participation of the University of Michigan's Operations and Industrial Engineering department and the Mathematics department with the CAARMS5 meeting. We also thank the National Security Agency, the University of Michigan, and Morgan State University for their financial support.

Finally we express our deep gratitude to all the anonymous referees and to Christine Thivierge of the American Mathematical Society for their advice and comments.

Alfred G. Noël
University of Massachusetts Boston

Earl Barnes
Georgia Institute of Technology

Sonya A. F. Stephens
Florida A & M University

December 2000

I. Research and Expository Papers

Contemporary Mathematics
Volume **275**, 2001

A Lower Bound for the Chromatic Number of a Graph

Earl R. Barnes

December 15, 2000

School of Industrial and Systems Engineering
Georgia Institute of Technology
Atlanta, Georgia 30332-0205

Abstract

Hoffman [1] has given a lower bound on the chromatic number of a graph in terms of the extreme eigenvalues of the adjacency matrix of the graph. In this paper we give a sharper lower bound on the chromatic number. Our bound requires the solution of a semidefinite programming problem which depends on an extreme eigenvector of the adjacency matrix. We describe an efficient algorithm for solving this problem.

1 Introduction.

Graph coloring problems arise in the scheduling of exams for university courses [2], the scheduling of municipal services [3], the testing of computer circuits [4], and in many other areas. We will give a small example of the exam scheduling problem to motivate our dicussion of the graph coloring problem.

Imagine a small university which has 10 students $s_1, \ldots, s_{10}$, and offers 7 courses $c_1, \ldots, c_7$. Each student is registered in a subset of the as indicated in the following table.

2000 *Mathematics Subject Classifications.* Primary 05C15, 05C85, 15A18, 90C22.

$s_1 \in \{c_1, c_2\}$	$s_6 \in \{c_6, c_7\}$
$s_2 \in \{c_1, c_3, c_4\}$	$s_7 \in \{c_2, c_4\}$
$s_3 \in \{c_4, c_7\}$	$s_8 \in \{c_5, c_6, c_7\}$
$s_4 \in \{c_2, c_3, c_4\}$	$s_9 \in \{c_2, c_6\}$
$s_5 \in \{c_1, c_2\}$	$s_{10} \in \{c_3, c_5\}$

Table 1

The problem is to schedule exams for all the courses in such a way that no student has an exam conflict, and the number of examination periods required is as small as possible. To abstract the essence of this problem we construct a graph in which each class is a node. We then connect two nodes with an edge if there is a single student registered in both of these courses. The resulting graph is shown in Fig. 1.

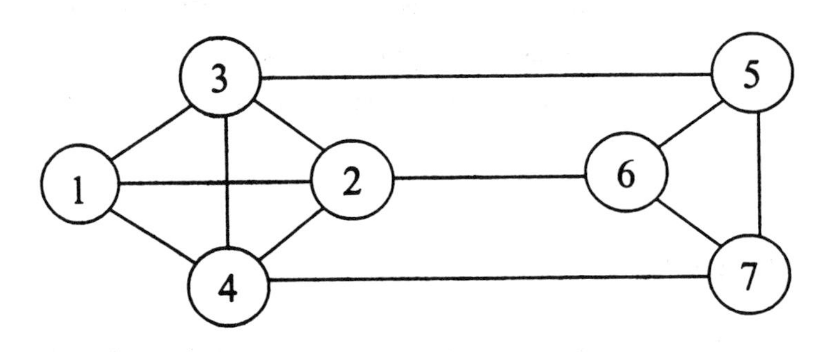

Conflict Graph

Figure 1

Let the exam periods be represented by colors. The problem then is to determine the least number of colors needed to color the nodes of the graph in such a way that if two nodes are connected they get different colors. An assignment of colors to nodes is equivalent to an assignment of courses to exam periods. We can color the nodes of the graph in Fig. 1 using four colors. Clearly this is a minimum.

Definition 1.1 The *chromatic number* of a graph is the least number of colors needed to color the nodes of the graph in such a way that two nodes get different colors if they are connected.

Let G be a graph on n nodes $\{1,\ldots,n\}$ and let χ denote the chromatic number of G. Throughout what follows we assume G is connected. In general it is NP-hard to compute χ, but bounds on χ are easy to compute. For example, let $A=(a_{ij})$ denote the adjacency matrix of G. Thus

$$a_{ij}=\begin{cases}1 & \text{if nodes } i \text{ and } j \text{ are connected}\\ 0 & \text{otherwise}\end{cases}$$

Let $\lambda_1 \geq \lambda_2 \geq \cdots \geq \lambda_n$ denote the eigenvalues of A. In [5] Wilf shows that $\lambda_1+1\geq\chi$. In [1] Hoffman shows that

$$\chi \geq 1+\frac{\lambda_1}{|\lambda_n|}. \tag{1.1}$$

In the next Section we obtain a sharper lower bound for χ.

Example 1. The adjacency matrix for the graph in Fig. 1 is

$$A=\begin{bmatrix}0&1&1&1&0&0&0\\1&0&1&1&0&1&0\\1&1&0&1&1&0&0\\1&1&1&0&0&0&1\\0&0&1&0&0&1&1\\0&1&0&0&1&0&1\\0&0&0&1&1&1&0\end{bmatrix}.$$

It's extreme eigenvalues are $\lambda_1=3.5141$ and $\lambda_7=-2$. Wilf's theorem gives the upper bound $1+\lambda_1=4.5141$ for χ and Hoffman's theorem gives the lower bound $1+\lambda_1/|\lambda_7|=2.7571$.

2 Lower Bounds for χ.

The following theorem gives a generalization of (1.1).

Theorem 2.1 *Let A denote the adjacency matrix for a connected graph G on n nodes, and let $D = diag\ (d_1, \ldots, d_n)$ be a diagonal matrix such that $A + D$ is positive semidefinite. Then each d_i is positive, and the largest eigenvalue of the matrix $D^{-\frac{1}{2}}AD^{-\frac{1}{2}} + I$ is a lower bound for χ.*

Proof. Since $A+D$ is positive semidefinite each diagonal term d_i, and each principal submatrix of $A + D$, is positive semidefinite. Suppose that some d_i is 0. Since G is connected there is a node j such that $a_{ij} = 1$. The principal submatrix

$$\begin{pmatrix} d_i & a_{ij} \\ a_{ji} & d_j \end{pmatrix}$$

has determinant -1, and consequently is not positive semidefinite. But this contradicts the fact that $A + D$ is positive semidefinite. It follows that $d_i > 0$ for each i, and the matrix $D^{-\frac{1}{2}}$ is well defined.

To show that the largest eigenvalue of $D^{-\frac{1}{2}}AD^{-\frac{1}{2}} + I$ is $\leq \chi$ it suffices to show that

$$x^T(D^{-\frac{1}{2}}AD^{-\frac{1}{2}} + I)x \leq \chi \tag{2.1}$$

for each unit vector x. In general we will denote the largest eigenvalue of a real symmetric matrix M by $\lambda_{\max}(M)$. Observe that

$$A + D = D^{\frac{1}{2}}(D^{-\frac{1}{2}}AD^{-\frac{1}{2}} + I)D^{\frac{1}{2}}$$

Thus the matrix $M = D^{-\frac{1}{2}}AD^{-\frac{1}{2}} + I$ is positive semidefinite. We can therefore find an $n \times n$ matrix U such that $U^TU = M$. Let $u_1, \ldots, u_n$ denote the columns of U. We then have

$$u_i^T u_j = \begin{Bmatrix} \frac{a_{ij}}{\sqrt{d_i d_j}} & \text{if } i \neq j \\ 1 & \text{if } i = j \end{Bmatrix} \tag{2.2}$$

Let $C_1, \ldots, C_\chi$ denote the color classes in a proper coloring of G. By (2.2) the set of vectors $\{u_i | i \in C_k\}$ is orthonormal for each $k = 1, \ldots, \chi$. This means that for any unit vector x we have

$$\sum_{i \in C_k} (u_i^T x)^2 \leq \|x\|^2 = 1.$$

It follows that

$$x^T(UU^T)x = x^T(\sum_{i=1}^{n} u_i u_i^T)x = \sum_{i=1}^{n}(u_i^T x)^2$$

$$= \sum_{k=1}^{\chi}\sum_{i\in C_k}(u_i^T x)^2 \leq \sum_{k=1}^{\chi} 1 = \chi.$$

This shows that

$$\lambda_{\max}(UU^T) = \max_{|x|=1} x^T(UU^T)x \leq \chi.$$

However,

$$\lambda_{\max}(UU^T) = \lambda_{\max}(U^TU) = \lambda_{\max}(D^{-\frac{1}{2}}AD^{-\frac{1}{2}} + I) \leq \chi. \tag{2.3}$$

This completes the proof of the theorem.

All that remains now is to describe a procedure for computing a diagonal matrix D such that $A + D$ is positive semidefinite and the bound (2.3) is interesting. We will do this in the next section. First observe that Hoffman's bound (1.1) can be obtained from (2.3) by taking $D = |\lambda_n|I$ in Theorem 2.1.

3 Computing the Matrix D.

We wish to determine a matrix D for which inequality (2.3) gives a bound on χ which is at least as good as the bound (1.1). For this we have the following theorem.

Theorem 3.1 *Let $v = (v_1, \ldots, v_n)^T$ denote the unit eigenvector of A corresponding to the eigenvalue $\lambda_{\max}(A)$ and let $d_1, \ldots, d_n$ be a solution for the following semidefinite programming problem.*

$$\text{minimize } v_1^2 d_1 + v_2^2 d_2 + \cdots + v_n^2 d_n$$

$$\text{subject to } A + D \geq 0, \text{ where } D = diag(d_1, \ldots, d_n). \tag{3.1}$$

Then

$$1 + \frac{\lambda_1}{|\lambda_n|} \leq \lambda_{\max}(D^{-\frac{1}{2}}AD^{-\frac{1}{2}} + I) \leq \chi. \tag{3.2}$$

Proof. Since $A + |\lambda_n|I \geq 0$ and $\|v\|^2 = 1$ we have

$$|\lambda_n| = \sum_{i=1}^{n} v_i^2 |\lambda_n| \geq \sum_{i=1}^{n} v_i^2 d_i = (D^{\frac{1}{2}} v)^T (D^{\frac{1}{2}} v).$$

It follows that

$$\frac{\lambda_1}{|\lambda_n|} \leq \frac{v^T A v}{(D^{\frac{1}{2}} v)^T (D^{\frac{1}{2}} v)} = \frac{(D^{\frac{1}{2}} v)^T (D^{-\frac{1}{2}} A D^{-\frac{1}{2}})(D^{\frac{1}{2}} v)}{\|D^{\frac{1}{2}} v\|^2} = u^T (D^{-\frac{1}{2}} A D^{-\frac{1}{2}}) u$$

where

$$u = \frac{D^{\frac{1}{2}} v}{\|D^{\frac{1}{2}} v\|}$$

is a unit vector. Now observe that

$$1 + \frac{\lambda_1}{|\lambda_n|} \leq u^T (D^{-\frac{1}{2}} A D^{-\frac{1}{2}} + I) u \leq \max_{\|u\|=1} u^T (D^{-\frac{1}{2}} A D^{-\frac{1}{2}} + I) u = \lambda_{\max}(D^{-\frac{1}{2}} A D^{-\frac{1}{2}} + I).$$

This completes the proof of the first inequality in (3.2). The second inequality is just a restatement of (2.3).

4 The Semidefinite Programming Problem (3.1).

The semidefinite programming problem (3.1) is equivalent to the following linear programming problem which has an infinite number of constraints.

$$\begin{aligned} &\text{minimize} && v_1^2 d_1 + v_2^2 d_2 + \cdots + v_n^2 d_n \\ &\text{subject to} && x^T (A + D) x \geq 0, x \in S^{n-1} \end{aligned} \tag{4.1}$$

where S^{n-1} denotes the unit sphere in E^n. We will describe a procedure which can be used to solve this problem to any desired degree of accuracy.

Let $\varepsilon > 0$ be given. Let $C = \{x_1, \ldots, x_N\}$ be a set of vectors in S^{n-1}. The linear programming problem (2.6) can be approximated by the following problem which has only a finite number of constraints.

$$\begin{aligned} &\text{minimize} && v_1^2 d_1 + v_2^2 + \cdots + v_n^2 d_n \\ &\text{subject to} && x_i^T (A + D) x_i \geq \varepsilon, x_i \in C. \end{aligned} \tag{4.2}$$

We will show how to choose C so that the value of the minimum in (4.2) is within ε of the value of the minimum in (4.1).

Lemma 3.1. If the solution $D^* = (d_1^*, \dots, d_n^*)$ of (4.2) satisfies $A + D^* \geq 0$, then the value of the minimum in (4.2) differs by less than ε from the value of the minimum in (3.1).

Proof. Let $\overline{v}$ denote the value of the minimum in (4.2) and let $\underline{v}$ denote the value of the minimum in (3.1). Since $A + D^* \geq 0$, D^* is feasible for (4.1). It follows that $\underline{v} \leq \overline{v}$. We will show that $\overline{v} \leq \underline{v} + \varepsilon$. The dual of the semidefinite programming problem (3.1) is

$$\begin{array}{ll} \text{maximize} & -\text{ trace } (AX) \\ \text{subject to} & \text{trace } (A_i X) = v_i^2, i = 1, \dots, n, \\ & X \geq 0, \end{array} \tag{4.3}$$

where A_i is an $n \times n$ matrix with a one in position (i,i) and zeros elsewhere. The value of this maximum is of course $\underline{v}$.

Let x_{ji} denote the jth component of the vector x_i. (4.2) can then be written as

$$\begin{array}{ll} \text{minimize} & v_1^2 d_1 + v_2^2 d_2 + \cdots + v_n^2 d_n \\ \text{subject to} & x_{11}^2 d_1 + x_{21}^2 d_2 + \cdots + x_{n1}^2 d_n \geq \varepsilon - x_1^T A x_1 \\ & x_{12}^2 d_1 + x_{22}^2 d_2 + \cdots + x_{n2}^2 d_n \geq \varepsilon - x_2^T A x_2 \\ & \dots \\ & x_{1N}^2 d_1 + x_{2N}^2 d_2 + \cdots + x_{nN}^2 d_n \geq \varepsilon - x_N^T A x_N. \end{array} \tag{4.4}$$

The value of this minimum is $\overline{v}$.

The dual of (4.4) is

$$\begin{array}{ll} \text{maximize} & y_1(\varepsilon - x_1^T A x_1) + \cdots + y_N(\varepsilon - x_N^T A x_N) \\ \text{subject to} & y_1 x_{11}^2 + y_2 x_{12}^2 + \cdots + y_N x_{1N}^2 = v_1^2 \\ & y_1 x_{21}^2 + y_2 x_{22} + \cdots + y_N x_{2N} = v_2^2 \\ & \dots \\ & y_1 x_{n1}^2 + y_2 x_{n2}^2 + \cdots + y_N x_{nN} = v_n^2 \\ & y_1, \dots, y_N \geq 0. \end{array} \tag{4.5}$$

The value of this maximum is of course $\overline{v}$.

Let $y_1, \dots, y_N$ be a solution of (4.5) and define

$$X^* = y_1 x_1 x_1^T + \cdots + y_N x_N x_N^T.$$

Clearly $X^* \geq 0$, and the constraints in (4.5) say that

$$\text{trace } (A_i X^*) = v_i^2, i = 1, \ldots, n.$$

Thus X^* is feasible for (4.3) and so

$$\underline{v} \geq - \text{ trace } (AX^*) = -(y_1 x_1^T A x_1 + \cdots + y_N x_N^T A x_N). \tag{4.6}$$

If we add the constraints in (4.5) and use the fact that $x_1, \ldots, x_N, v$ are unit vectors we see that $y_1 + \cdots + y_N = 1$. Now comparing (4.6) with the objective in (4.5) we see that

$$\underline{v} \geq - \text{ trace } (AX^*) = (\varepsilon - \text{ trace}(AX^*)) - \varepsilon = \overline{v} - \varepsilon.$$

This completes the proof of the lemma.

In order to complete the description of our algorithm for solving (3.1) we must explain how to compute vectors $x_1, \ldots, x_N$ satisfying the conditions of Lemma 3.1. This can be done as follows.

Step 0. Let $C = \{x_1, \ldots, x_n\}$ be a set of orthonormal vectors in E^n. Set $N = n$ and solve the linear programming problem (4.4). Denote the solution by D^*.

Step 1.

If $A + D^* \geq 0$ stop. The conditions of Lemma 3.1 are satisfied. Otherwise go to Step 2.

(4.7)

Step 2. Set $N = N + 1$ and find a unit vector x_N satisfying $x_N^T(A + D^*)x_N < 0$. Add x_N to C and solve (4.4). Denote the solution by D^* and return to Step 1.

The Gaussian elimination algorithm can be used to perform the test in Step 1. If the test fails the algorithm produces the vector x_N required in Step 2. For details see Proposition 2.4.3 in [6].

Theorem 4.2 *The procedure described in (4.7) terminates for a finite value of N.*

Proof. From the proof of Theorem 3.1 we see that $v_1^2 d_1^* + \cdots + v_n^2 d_n^* \leq |\lambda_n|$ for each D^*. The Perron eigenvector v is strictly positive. Thus there exists a constant

κ such that $d_i^* \leq \kappa, i = 1, \ldots, n$ for each D^*. Consider any x_N computed in Step 2. By the Mean Value-Theorem of differential calculus we have, for any $j < N$,

$$\varepsilon \leq x_j^T(A + D^*)x_j - x_N^T(A + D^*)x_N$$

$$= 2((A + D^*)z)^T(x_j - x_N),$$

where z is a convex combination of the unit vectors x_j and x_N. It follows that $\|z\| \leq 1$ and therefore

$$\varepsilon \leq \|A + D^*\| \|x_j - x_N\| \leq (\lambda_1 + \kappa)\|x_j - x_N\|.$$

This shows that the distance between any two of the points $\{x_1, \ldots, x_N\}$ is at least $\min\{1, \frac{\varepsilon}{\lambda_1+\kappa}\}$ Since S^{n-1} is compact the sequence generated by (4.7) must therefore terminate for some finite value of N. This completes the proof of the theorem.

Example 1 (continued). For the graph from our exam scheduling problem the procedure (4.7) with $\varepsilon = .1$ gives

$$D = \text{ diag } (.7868, 2.5534, 2.4002, 2.1292, 1.7998, 1.6466, 2.0708).$$

The largest eigenvalue of the matrix $D^{-\frac{1}{2}}AD^{-\frac{1}{2}} + I$ is 3.9521. This is a better lower bound for χ than Hoffman's bound.

References

[1] A. J. Hoffman, "On eigenvalues and colorings of graphs," Graph Theory and its Applications (B. Harris, ed.), Academic Press, 1970, pp. 78-91.

[2] M. W. Carter, G Laporte, and J. W. Chinneck, "A general examination scheduling system," Interfaces, Vol. 24, No. 3, 1994, pp. 109-120.

[3] A. Tucker, Applied Combinatorics, John Wiley and Sons, 1995,Section 2.3.

[4] M. R. Garey, D. S. Johnson, and H. C. So,"An application of graph coloring to printed circuit testing," IEEE Trans. Circuits and Systems CAS, 1976, pp.591-599.

[5] H. S. Wilf, "The eigenvalues of a graph and its chromatic number," J. London Math. Society, Vol. 42, 1967, pp. 330-332.

[6] M. M. Deza and M. Laurent, Geometry of Cuts and Metrics, Springer Verlag, 1997.

Contemporary Mathematics
Volume **275**, 2001

THE FRACTIONAL PARTS OF $\frac{N}{K}$

M. R. CURRIE AND E. H. GOINS, R21

National Security Agency and Stanford University

ABSTRACT. Let $[\cdot]$ denote the greatest integer function and $\{x\} = x - [x]$. Then

$$\lim_{n\to\infty} \frac{1}{n} \sum_{k=1}^{n} \left\{ \frac{n}{k} \right\} = 1 - \gamma,$$

where $\gamma = 0.57721566\ldots$ is the Euler-Mascheroni constant. We generalize to results for the greatest multiple of $\frac{1}{m}$ and nearest multiple of $\frac{1}{m}$. In addition, we examine the limiting distribution of the fractional parts, establishing along the way connections with Stirling's approximation as well as the gamma and digamma functions. We also show that the distribution of "shifts" of the fractional parts approaches uniformity. Finally, we examine the idiosyncrasies of a measure induced by the distribution of the fractional parts.

A Quandary

Proposition 1. *Let $[\cdot]$ denote the greatest integer function and $\{x\} = x - [x]$. Then*

$$\lim_{n\to\infty} \frac{1}{n} \sum_{k=1}^{n} \left\{ \frac{n}{k} \right\} = 1 - \gamma,$$

where γ is the Euler-Mascheroni constant defined by

$$\gamma = \lim_{n\to\infty} \left(\sum_{k=1}^{n} \frac{1}{k} - \log n \right).$$

This proposition is counterintuitive. In our innocence, we would have been comforted had the limit been $\frac{1}{2}$. We give a proof to the contrary. Before launching into it, we note that this proposition can be established by modifying the classical proof of Dirichlet's asymptotic formula for the partial sums of the divisor function. See [1] and [5] for more on this subject. However, the technique we present below generalizes nicely and is typical of the approach we take in the rest of the paper.

Key words and phrases. Euler-Mascheroni constant, Stirling's approximation, mod 1, gamma function, digamma function, measure, continued fraction, uniformity, distribution.

2000 *Mathematics Subject Classifications.* Primary 11K06.

(Proof) The key is to view our average as a Riemann sum and its limit as the corresponding Riemann integral.[1]

$$
\begin{aligned}
&\lim_{n\to\infty} \frac{1}{n}\sum_{k=1}^{n}\left\{\frac{n}{k}\right\} \\
&= \int_0^1 \left(\frac{1}{x} - \left[\frac{1}{x}\right]\right) dx = \int_1^\infty \frac{x-[x]}{x^2}\,dx = \lim_{n\to\infty}\sum_{k=1}^{n-1}\int_k^{k+1}\frac{x-k}{x^2}\,dx \\
&= \lim_{n\to\infty}\sum_{k=1}^{n-1}\left(\log x + \frac{k}{x}\right)\Big|_k^{k+1} = \lim_{n\to\infty}\left(\log n - \sum_{k=2}^{n}\frac{1}{k}\right) \\
&= 1-\gamma.
\end{aligned}
$$

One might conjecture that divisors play a role in producing an average that is not equal to $1/2$, which we had intuitively thought it should be. It is easy to dispel this notion by restricting our considerations to the sequence of primes. The limit is, of course, the same for this subsequence. It is clear then that divisors do not account for this phenomenon.

However, it may be helpful to reflect on the fact that we have measured the average value of $\frac{1}{x} - [\frac{1}{x}]$ on the interval $[0,1]$. It is easy to see that the average value of this function on the interval $[1/2, 1]$ is $\log 4 - 1$. Generally, on the interval $[\frac{1}{j+1}, \frac{1}{j}]$, the average value is

$$\frac{\log\left(1+\frac{1}{j}\right) - \frac{1}{j+1}}{\frac{1}{j} - \frac{1}{j+1}}.$$

One application of L'Hôpital's rule shows that this expression goes to $1/2$ as j goes to infinity. So if we average $\{\frac{n}{k}\}$ only over those values of k for which $\frac{n}{k}$ is "large", our average will indeed be close to the value $1/2$ that we had anticipated.

We can cast this problem in another light by considering the continued fraction expansion $[0; a_1, a_2, a_3, \ldots]$ for real numbers in $[0,1]$. Proposition 1. says that the average value of $[0; a_2, a_3, a_4, \ldots]$ is $1-\gamma$. So our proposition is tied to Gauss' problem and Kuz'min's theorem. For more on this see [6].

We now generalize Proposition 1. One may observe that our definition below is equivalent to saying that $[x]_m = [mx]/m$. We introduce the new notation for convenience. In particular, it allows us to bridge easily to the notation we need for the nearest multiple function, as well as easing the discussion in our section on measure-theoretical aspects. Writing $\{x\}_m = x - [x]_m$, we have the following proposition.

[1] The reader may feel somewhat squeamish about the first step here. It can be justified by noting that the function $f(x) = \frac{1}{x} - \left[\frac{1}{x}\right]$ is bounded and continuous except on a countable set, so f is Riemann integrable. Although f is not defined at zero, this issue is addressed by defining $f(0)$ to be 0. At this point we also note for future reference that for any piecewise continuous function g, $g(f)$ is bounded and continuous a.e., hence Riemann integrable. (In fact $g(f)$ will have only countably many discontinuities.) We use this line of reasoning throughout the paper.

Proposition 2. *Given a positive integer m, let $[x]_m$ denote the greatest multiple of $1/m$ less than or equal to x. Then we have the limit*

$$\lim_{n\to\infty}\frac{1}{n}\sum_{k=1}^{n}\left\{\frac{n}{k}\right\}_m=\sum_{k=1}^{m}\frac{1}{k}-\log m-\gamma.$$

(Proof) Using the same idea as that above, we have

$$\lim_{n\to\infty}\frac{1}{n}\sum_{k=1}^{n}\left\{\frac{n}{k}\right\}_m=\int_1^\infty\frac{x-[x]_m}{x^2}\,dx$$

$$=\lim_{n\to\infty}\sum_{k=1}^{n-1}\sum_{\ell=0}^{m-1}\int_{k+\frac{\ell}{m}}^{k+\frac{\ell+1}{m}}\frac{x-\left(k+\frac{\ell}{m}\right)}{x^2}\,dx=\lim_{n\to\infty}\sum_{k=1}^{n-1}\sum_{\ell=0}^{m-1}\left(\log x+\frac{k+\frac{\ell}{m}}{x}\right)\Bigg|_{k+\frac{\ell}{m}}^{k+\frac{\ell+1}{m}}$$

Employing the identity

$$\left(\frac{k+\frac{\ell}{m}}{x}\right)\Bigg|_{k+\frac{\ell}{m}}^{k+\frac{\ell+1}{m}}=\left(k+\frac{\ell}{m}\right)\left(\frac{1}{k+\frac{\ell+1}{m}}-\frac{1}{k+\frac{\ell}{m}}\right)=-\frac{1}{mk+\ell+1}$$

yields

$$\lim_{n\to\infty}\frac{1}{n}\sum_{k=1}^{n}\left\{\frac{n}{k}\right\}_m=\lim_{n\to\infty}\left(\log n-\sum_{k=1}^{n-1}\sum_{\ell=0}^{m-1}\frac{1}{mk+\ell+1}\right)$$

$$=\lim_{n\to\infty}\left(\log n-\sum_{k=m+1}^{mn}\frac{1}{k}\right)=\lim_{n\to\infty}\left(\sum_{k=1}^{m}\frac{1}{k}-\log m+\log mn-\sum_{k=1}^{mn}\frac{1}{k}\right)$$

$$=\sum_{k=1}^{m}\frac{1}{k}-\log m-\gamma.$$

Proposition 1. now has an interesting context. It is the error found when approximating the constant γ by the sum $\sum_1^m 1/k-\log m$, for the case $m=1$.

We omit the proof of the following proposition, since it is quite similar to that of Proposition 2.

Proposition 3. *Given a positive integer m, let $\langle x\rangle_m$ denote the multiple of $1/m$ nearest*[2] *x. Then we have*

$$\lim_{n\to\infty}\frac{1}{n}\sum_{k=1}^{n}\left(\frac{n}{k}-\left\langle\frac{n}{k}\right\rangle_m\right)=\sum_{k=1}^{m}\frac{2}{2k-1}-\log 4m-\gamma.$$

[2]For the record, we define $\left\langle\frac{k}{m}+\frac{1}{2m}\right\rangle_m$ to be $\frac{k+1}{m}$.

Again, this is an error in estimating the Euler-Mascheroni constant. This theme will be repeated when we examine the variance, where we find that errors in estimating other well known limits appear. However, we first discuss the distribution of fractional parts, in the limit.

Probability Distributions

We want to know how many fractional parts $\left\{\frac{n}{k}\right\}_m$ are in a given interval. To be somewhat more precise, we wish to calculate a limiting distribution $\mathcal{P}_m$.

For any $I = [a, b] \subseteq [0, \frac{1}{m}]$, let

$$\omega_I(x) = \begin{cases} 1, & \text{when } x \in I; \\ 0, & \text{otherwise.} \end{cases}$$

and

$$\mathcal{P}_m(I) = \lim_{n\to\infty} \frac{1}{n} \sum_{k=1}^{n} \omega_I \left\{ \frac{n}{k} \right\}_m .$$

Then

$$\begin{aligned} \mathcal{P}_m(I) &= \lim_{n\to\infty} \sum_{k=1}^{n-1} \sum_{\ell=0}^{m-1} \int_{k+\frac{\ell}{m}}^{k+\frac{\ell+1}{m}} \frac{\omega_I(x - [x]_m)}{x^2}\, dx = \lim_{n\to\infty} \sum_{k=1}^{n-1} \sum_{\ell=0}^{m-1} \int_{k+\frac{\ell}{m}+a}^{k+\frac{\ell}{m}+b} \frac{1}{x^2}\, dx \\ &= \lim_{n\to\infty} \sum_{k=!}^{n-1} \sum_{\ell=0}^{m-1} \left(\frac{1}{k+\frac{\ell}{m}+a} - \frac{1}{k+\frac{\ell}{m}+b} \right) = \sum_{k=m}^{\infty} \frac{m^2(b-a)}{(k+ma)(k+mb)}. \end{aligned}$$

We also express this in integral form.

$$\begin{aligned} \sum_{k=m}^{\infty} \frac{m^2(b-a)}{(k+ma)(k+mb)} &= \sum_{k=m}^{\infty} \left(\frac{m}{k+ma} - \frac{m}{k+mb} \right) \\ &= \sum_{k=m}^{\infty} \int_0^1 m \left(x^{k+ma-1} - x^{k+mb-1} \right) dx = \sum_{k=0}^{\infty} \int_0^1 x^k \left(x^{ma} - x^{mb} \right) m x^{m-1}\, dx. \end{aligned}$$

Then letting $u = x^m$ we get

$$\int_0^1 \left(\sum_{k=0}^{\infty} u^{k/m} \right) \left(u^a - u^b \right) du = \int_0^1 \frac{u^b - u^a}{u^{1/m} - 1}\, du.$$

This is summarized in Proposition 4.

Proposition 4. *Given a positive integer m, let $[a,b] \subseteq \left[0, \frac{1}{m}\right]$, and $\mathcal{P}_m([a,b])$ denote the proportion of fractional parts $\left\{\frac{n}{k}\right\}_m$ in the interval $[a,b]$, in the limit. Then*

$$\mathcal{P}_m([a,b]) = \sum_{k=m}^{\infty} \frac{m^2(b-a)}{(k+ma)(k+mb)} = \int_0^1 \frac{x^b - x^a}{x^{1/m}-1}\, dx.$$

In particular, using the integral representation it is a straightforward matter to see that for $m = 1$

$$\mathcal{P}_1([1/2, 1]) = \log 4 - 1.$$

It is perhaps interesting to recall that $\log 4 - 1$ is also the average value of $\frac{1}{x} - \left[\frac{1}{x}\right]$ on the interval $\left[\frac{1}{2}, 1\right]$.

Probability Densities

Using the expressions for the probability distribution we found above, we derive the probability density functions.

Proposition 5. *Let $\theta \in [0,1]$ and $\mathcal{P}_1(I)$ be the probability distribution as defined above. Then the probability density function $\mathcal{D}_1$ is given by*

$$\mathcal{D}_1(\theta) = \lim_{h\to 0} \frac{\mathcal{P}_1([\theta, \theta+h])}{h} = \sum_{k=1}^{\infty} \frac{1}{(k+\theta)^2} = \int_0^1 \frac{\log x}{x-1}\, x^\theta \, dx.$$

It is easy to see that the density function is monotone decreasing and one notes that the density at zero is $\frac{\pi^2}{6}$. Furthermore, the second derivative is always positive, so the density function is concave up.

We calculate the density functions for the general case as well.

Proposition 6. *Given a positive integer m, let $\theta \in \left[0, \frac{1}{m}\right]$, and $\mathcal{P}_m(I)$ be the probability distribution function as defined above. Then the probability density function $\mathcal{D}_m$ is given by*

$$\mathcal{D}_m(\theta) = \lim_{h\to 0} \frac{\mathcal{P}_m([\theta, \theta+h])}{h} = \sum_{k=m}^{\infty} \frac{m^2}{(k+m\theta)^2} = \int_0^1 \frac{\log x}{x^{1/m}-1}\, x^\theta \, dx.$$

It should be noted here that the density function for $\langle\cdot\rangle_m$ is easily derived from the density function for $[\cdot]_m$. In fact, it is the same as the density function for $[\cdot]_m$ on $[1, \frac{1}{2m}]$ and the density function on $[-\frac{1}{2m}, 0]$ is a translation from $[\frac{1}{2m}, \frac{1}{m}]$ of $\mathcal{D}_m$.

Calculating the Variance and Mean Absolute Deviation

We wish to calculate the variance of the limiting distribution $\mathcal{P}_m$ established in the previous section. To facilitate this we first calculate the second moment.

Proposition 7. *For each positive integer m we have*

$$\lim_{n\to\infty} \frac{1}{n} \sum_{k=1}^{n} \left\{ \frac{n}{k} \right\}_m^2 = \frac{1}{m} \left(\sum_{k=1}^{m} \frac{1}{k} - \log m - \gamma \right) - \frac{2}{m} \log \left(\frac{m!}{\sqrt{2\pi}\; m^{m+\frac{1}{2}}\; e^{-m}} \right).$$

(Proof) The calculation begins as usual

$$\begin{aligned}
\lim_{n\to\infty} \frac{1}{n} \sum_{k=1}^{n} \left\{ \frac{n}{k} \right\}_m^2
&= \int_1^\infty \frac{(x-[x]_m)^2}{x^2}\, dx = \lim_{n\to\infty} \sum_{k=1}^{n-1} \sum_{\ell=0}^{m-1} \int_{k+\frac{\ell}{m}}^{k+\frac{\ell+1}{m}} \left(1 - 2\,\frac{[x]_m}{x} + \frac{([x]_m)^2}{x^2} \right) dx \\
&= \lim_{n\to\infty} \sum_{k=1}^{n-1} \sum_{\ell=0}^{m-1} \int_{k+\frac{\ell}{m}}^{k+\frac{\ell+1}{m}} \left(1 - 2\,\frac{k+\frac{\ell}{m}}{x} + \frac{(k+\frac{\ell}{m})^2}{x^2} \right) dx.
\end{aligned}$$

We simplify this expression by using the identity

$$\int_{k+\frac{\ell}{m}}^{k+\frac{\ell+1}{m}} \left(1 + \frac{(k+\frac{\ell}{m})^2}{x^2} \right) dx = \frac{1}{m} + \frac{k+\frac{\ell}{m}}{mk+\ell+1} = \frac{1}{m}\left(2 - \frac{1}{mk+\ell+1} \right),$$

as well as the expression

$$\begin{aligned}
\int_{k+\frac{\ell}{m}}^{k+\frac{\ell+1}{m}} \left(-2\,\frac{k+\frac{\ell}{m}}{x} \right) dx &= -2\left(k + \frac{\ell}{m} \right) \log \frac{mk+\ell+1}{mk+\ell} \\
&= \frac{2}{m} \log(mk+\ell+1) \\
&\quad - \frac{2}{m} \left((mk+\ell+1)\log(mk+\ell+1) - (mk+\ell)\log(mk+\ell) \right).
\end{aligned}$$

The latter expression gives a telescoping sum, so the only terms that do not vanish are at the endpoints: $k=1$, $\ell=0$; and $k=n-1$, $\ell=m-1$. The entire expression then simplifies to

$$\sum_{k=1}^{n-1}\sum_{\ell=0}^{m-1}\int_{k+\frac{\ell}{m}}^{k+\frac{\ell+1}{m}}\left(1-2\frac{k+\frac{\ell}{m}}{x}+\frac{(k+\frac{\ell}{m})^2}{x^2}\right)dx$$
$$=\frac{1}{m}\left(2m(n-1)-\sum_{k=m+1}^{mn}\frac{1}{k}\right)-\frac{2}{m}(mn\log mn-m\log m)+\frac{2}{m}\log\frac{(mn)!}{m!}$$
$$=\frac{1}{m}\left(\log mn-\sum_{k=m+1}^{mn}\frac{1}{k}-\log m\right)$$
$$-\frac{2}{m}\left(m-(m+\tfrac{1}{2})\log m+\log m!-\left(mn-(mn+\tfrac{1}{2})\log mn+\log(mn)!\right)\right).$$

As n increases without bound, the first term tends to a familiar value. It is $1/m$ times the error term found when approximating the Euler-Mascheroni constant.

$$\lim_{n\to\infty}\frac{1}{m}\left(\log mn-\sum_{k=m+1}^{mn}\frac{1}{k}-\log m\right)=\frac{1}{m}\left(\sum_{k=1}^{m}\frac{1}{k}-\log m-\gamma\right).$$

The treatment for the second term is less obvious, but it does tend to a limit. Recall that according to the limit attributed to Stirling, known as Stirling's approximation, we have:

$$\lim_{k\to\infty}\log\frac{k!}{k^{k+\frac{1}{2}}e^{-k}}=\log\sqrt{2\pi}.$$

We then have the limiting value

$$\lim_{n\to\infty}\left(mn-(mn+\tfrac{1}{2})\log mn+\log(mn)!\right)=\lim_{mn\to\infty}\log\left(\frac{(mn)!}{(mn)^{mn+\frac{1}{2}}\,e^{-mn}}\right)=\log\sqrt{2\pi}.$$

Note that the second moment is a linear combination of the logarithm of error terms; the first approximating γ, the second essentially approximating $m!$. Note also that this linear combination must always be positive, which is not obvious. The variance is now a direct consequence of this proposition.

Proposition 8. *Let the limit of the average value of the fractional parts be denoted by*

$$a_m=\lim_{n\to\infty}\frac{1}{n}\sum_{k=1}^{n}\left\{\frac{n}{k}\right\}_m=\left(\sum_{k=1}^{m}\frac{1}{k}-\log m\right)-\gamma.$$

Then the variance σ_m^2 of $\mathcal{P}_m$ satisfies the relation

$$\begin{aligned}\sigma_m^2 &= \lim_{n\to\infty}\sum_{k=1}^{n}\left(\left\{\frac{n}{k}\right\}_m - a_m\right)^2\frac{1}{n}\\ &=\frac{1}{m}\left(a_m - \log\left(\frac{m!}{\sqrt{2\pi}\; m^{m+\frac{1}{2}}\; e^{-m}}\right)^2\right) - a_m^2.\end{aligned}$$

For $m = 1$,

$$\sigma_1^2 = \gamma\,(1-\gamma) + \log 2\pi - 2.$$

Proposition 9. *If* a_m *is as above, then the mean absolute deviation from* a_m *for the limiting distribution* $\mathcal{P}_m$ *is given by*

$$2\log\frac{\Gamma(1+a_m m)}{\Gamma(1+a_m m, m)} + 2\log\frac{m}{(1+a_m)(1+a_m m+m)} + 2a_m(1-a_m m),$$

where

$$\Gamma(s) = \lim_{n\to\infty}\Gamma(s,n)$$

and

$$\Gamma(s,n) \equiv \frac{n!\,n^s}{s(1+s)(2+s)\dots(n+s)}.$$

For $m = 1$, *we have mean deviation*

$$2\log\Gamma\,(2-\gamma) + 2\gamma\,(1-\gamma)\,.$$

The derivation of Proposition 9 is a rather tedious exercise and we do not include it here. We remark on the interesting pattern of this formula and the appearance of yet another error term, this time involving the approximation of Γ.

An Observation

We now establish the rather pleasant fact that many of the results that we have presented may be expressed elegantly in terms of the digamma[3] function F. We give the definition as well as some of the function's properties here. For more background on the digamma function please see [2], [3] and [4].

$$F(x) \equiv \frac{d}{dx}(\log x!) = \frac{d}{dx}[\log \Gamma(x+1)] = \frac{\Gamma'(x+1)}{\Gamma(x+1)}.$$

It is a well known fact that the gamma function has the multiplicative property $\Gamma(x+1) = x\Gamma(x)$. This gives an identity for the digamma function:

$$F(x) = \frac{\Gamma(x) + x\Gamma'(x)}{x\Gamma(x)} = \frac{1}{x} + F(x-1).$$

A lesser known fact is that $F(0) = \Gamma'(1) = -\gamma$. Using this and the recursive property above, we find that for any positive integer m,

$$F(m) = \sum_{k=1}^{m} [F(k) - F(k-1)] + F(0) = \sum_{k=1}^{m} \frac{1}{k} - \gamma.$$

With this observation and Proposition 2. we get Proposition 10.

Proposition 10. *For all positive integers m we have*

$$\lim_{n\to\infty} \frac{1}{n} \sum_{k=1}^{n} \left\{ \frac{n}{k} \right\}_m = F(m) - \log m.$$

Hence, the result that first inspired us is $F(1)$. For arbitrary positive integral m, our results can be rewritten in terms of the digamma function.

Again the recursive property of the digamma function leads to the identity

$$F\left(k - \tfrac{1}{2}\right) = \frac{2}{2k-1} + F\left(k - \tfrac{3}{2}\right).$$

This, along with the fact that $F\left(-\frac{1}{2}\right) = -\log 4 - \gamma$, gives the relation

$$F\left(m - \tfrac{1}{2}\right) = \sum_{k=1}^{m} \left(F\left(k - \tfrac{1}{2}\right) - F\left(k - \tfrac{3}{2}\right)\right) + F\left(-\tfrac{1}{2}\right) = \sum_{k=1}^{m} \frac{2}{2k-1} - \log 4 - \gamma.$$

Using Proposition 3. and the discussion above we have:

---

[3]The reader may be acquainted with the polygamma function $\Psi(x) = \frac{d}{dx}[\log \Gamma(x)]$. We prefer a slightly different formulation for this application.

Proposition 11. *Let m be a positive integer. Then*

$$\lim_{n\to\infty}\frac{1}{n}\sum_{k=1}^{n}\left(\frac{n}{k}-\left\langle\frac{n}{k}\right\rangle_m\right)=\digamma\left(m-\tfrac{1}{2}\right)-\log m.$$

¿From Proposition 10. and 2., we see that $\digamma(m)-\log m\to 0$. In fact we need not restrict this to integral values. It is true that $\digamma(x)-\log x\to 0$. We demonstrate as follows.

$$|\digamma(x)-\log x|\leq|\digamma(x)-\digamma(m)|+|\digamma(m)-\log m|+|\log m-\log x|.$$

If we pick the integer m that satisfies $x\in[m,m+1]$, the monotonicity of *log* and $\digamma$ on $(0,\infty)$ gives us that the right-hand side of the inequality above is less than

$$|\digamma(m+1)-\digamma(m)|+|\digamma(m)-\log m|+|\log m-\log(m+1)|$$

$$\leq\frac{1}{m}+|\digamma(m)-\log m|+\log\frac{m+1}{m}\to 0.$$

Now,

$$\begin{aligned}\sum_{k=1}^{n}\frac{b-a}{(k+a)(k+b)}&=\sum_{k=1}^{n}\left(\frac{1}{k+a}-\frac{1}{k+b}\right)\\&=\sum_{k=1}^{n}\left(\digamma(k+a)-\digamma(k+a-1)-\digamma(k+b)+\digamma(k+b-1)\right)\\&=\digamma(b)-\digamma(a)+\\&\quad\digamma(n+a)-\log(n+a)+\log(n+b)-\digamma(n+b)+\log\frac{n+a}{n+b}.\end{aligned}$$

As n increases without bound, the last line goes to zero by the argument above. We have then the following intriguing result.

Proposition 12. *Let $[a,b]\subseteq[0,1]$, and $\mathcal{P}_1$ be the limiting distribution as defined earlier. Then*

$$\mathcal{P}_1([a,b])=\digamma(b)-\digamma(a).$$

The proof for the general case is completely analogous. We state the corresponding proposition below.

Proposition 13. *Given a positive integer m, let $[a,b]\subseteq\left[0,\frac{1}{m}\right]$. Then*

$$\mathcal{P}_m([a,b])=m\left(\digamma(mb+m-1)-\digamma(ma+m-1)\right).$$

Using Proposition 13., it is easy to calculate the probability density.

Proposition 14. *Given a positive integer m, let $\theta \in \left[0, \frac{1}{m}\right]$, the probability density function is given by*

$$\mathcal{D}_m(\theta) = \lim_{h \to 0} \frac{\mathcal{P}_m([\theta, \theta + h])}{h} = m^2 \mathcal{F}'(m\theta + m - 1).$$

The Average Value of the Tails

The developments in the preceding section cry out for a general statement regarding the digamma function and our averaging of sums associated with fractional parts. We attempt to address this issue with the following theorem.

Theorem 1. *Given a positive integer m and a function f bounded and continuous a.e. on [0,1], we have*

$$\lim_{n \to \infty} \frac{1}{n} \sum_{k=1}^{n} f\left\{\frac{n}{k}\right\}_m = \int_0^{\frac{1}{m}} m^2 f(\theta)\, \mathcal{F}'(m\theta + m - 1)\, d\theta.$$

In particular, when $m = 1$,

$$\lim_{n \to \infty} \frac{1}{n} \sum_{k=1}^{n} f\left(\left\{\frac{n}{k}\right\}\right) = \int_0^1 f(\theta)\, d\mathcal{F}(\theta).$$

(Proof) To begin the proof of this theorem, we note that

$$\lim_{n \to \infty} \frac{1}{n} \sum_{k=1}^{n} f\left\{\frac{n}{k}\right\}_m$$

$$= \int_1^{\infty} f(x - [x]_m) \frac{dx}{x^2} = \sum_{k=1}^{\infty} \sum_{\ell=0}^{m-1} \int_{k+\frac{\ell}{m}}^{k+\frac{\ell+1}{m}} f(x - [x]_m) \frac{dx}{x^2}.$$

On each subinterval in the integral, we have $[x]_m = k + \frac{\ell}{m}$. If we make the substitution $\theta = x - (k + \frac{\ell}{m})$, we find the simplified integral

$$\sum_{k=1}^{\infty} \sum_{\ell=0}^{m-1} \int_{k+\frac{\ell}{m}}^{k+\frac{\ell+1}{m}} f(x - [x]_m) \frac{dx}{x^2} = \sum_{k=1}^{\infty} \sum_{\ell=0}^{m-1} \int_0^{\frac{1}{m}} f(\theta) \frac{1}{\left(\theta + (k + \frac{\ell}{m})\right)^2} d\theta.$$

The theorem follows by using the identity below.

$$\sum_{k=1}^{\infty} \sum_{\ell=0}^{m-1} \frac{1}{\left(\theta + (k + \frac{\ell}{m})\right)^2} = \sum_{k=m}^{\infty} \frac{m^2}{(m\theta + k)^2}$$

$$= m^2 \mathcal{F}'(m\theta + m - 1).$$

The last equality holds by Propositions 6. and 14.

Many of the results in this paper follow as consequences of the theorem above; but it is a matter of integrating by parts and evaluating the digamma function, which means in general that the proofs are no simpler than those already given. However, we can use this machinery to examine a few new questions. What is the average value of the "tails" of the fractional parts we have been considering and what is the average value of the digit in the jth decimal place? We answer the first question in Proposition 15. It tells us that as we shift our fractional parts left and truncate the integral part, the average value converges to 1/2, which we might depict as a return to reason.

Proposition 15. *The average value of the tails converges to* $1/2$. *By this we mean*

$$\lim_{j\to\infty}\lim_{n\to\infty}\frac{1}{n}\sum_{k=1}^{n}\left\{10^j\left\{\frac{n}{k}\right\}\right\}=\frac{1}{2}.$$

(Proof) Since the function

$$f(x)=\{10^j x\}$$

is continuous except at those rational numbers that have both a terminating and infinite decimal expansion, we may use Theorem 1. to claim that evaluating the above limit is equivalent to evaluating the limit of the integral

$$\int_0^1\{10^j x\}dF(x)$$

as j goes to infinity. Now we have

$$\int_0^1\left(10^j x-[10^j x]\right)dF(x)=\sum_{k=0}^{10^j-1}\int_{\frac{k}{10^j}}^{\frac{k+1}{10^j}}\left(10^j x-k\right)dF(x).$$

If we do the integration by parts and simplify we get

$$\sum_{k=0}^{10^j-1}\int_{\frac{k}{10^j}}^{\frac{k+1}{10^j}}\left(10^j x-k\right)dF(x)=$$

$$\sum_{k=0}^{10^j-1}\left(F\left(\frac{k+1}{10^j}\right)-10^j\left(\log\Gamma\left(1+\frac{k+1}{10^j}\right)-\log\Gamma\left(1+\frac{k}{10^j}\right)\right)\right).$$

Note that the log terms telescope to

$$\log\Gamma(2)-\log\Gamma(1)=0.$$

So we are left with

$$\sum_{k=0}^{10^j-1} F\left(\frac{k+1}{10^j}\right) = \sum_{k=1}^{10^j} F\left(\frac{k}{10^j}\right),$$

which we want to evaluate as j goes to infinity. We may rewrite slightly to get 10^j times a Riemann sum for F on the interval [0,1].

$$\sum_{k=1}^{10^j} F\left(\frac{k}{10^j}\right) = 10^j\left(\frac{1}{10^j}\sum_{k=1}^{10^j} F\left(\frac{k}{10^j}\right)\right).$$

According to the trapezoidal rule, for partition of mesh w and trapezoidal approximation $T(w)$,

$$\int_a^b g(x)\,dx = T(w) + O(w^2).$$

We apply this to our situation and get

$$0 = \int_0^1 F(x)\,dx = T\left(\frac{1}{10^j}\right) + O\left(\frac{1}{10^{2j}}\right).$$

Recalling that F is monotone increasing on [0,1], we can replace $T\left(\frac{1}{10^j}\right)$ with our Riemann sum *minus* triangles of area

$$\frac{1}{2}\frac{1}{10^j}\left(F\left(\frac{k}{10^j}\right) - F\left(\frac{k-1}{10^j}\right)\right),$$

for each k. Since the sum of these triangles telescopes to $\frac{1}{2}\frac{1}{10^j}(F(1)-F(0)) = \frac{1}{2}\frac{1}{10^j}$, we can write

$$0 = \int_0^1 F(x)\,dx = \frac{1}{10^j}\sum_{k=1}^{10^j} F\left(\frac{k}{10^j}\right) - \frac{1}{2}\frac{1}{10^j} + O\left(\frac{1}{10^{2j}}\right).$$

Then

$$\frac{1}{2}\frac{1}{10^j} + O\left(\frac{1}{10^{2j}}\right) = \frac{1}{10^j}\sum_{k=1}^{10^j} F\left(\frac{k}{10^j}\right)$$

Multiplying on both sides by 10^j and taking the limit completes the proof.

We now turn to the question of the average value of the digit in the *jth* decimal place. Among other things, it follows from Proposition 16. that as j goes to infinity the average value converges to $4\frac{1}{2}$.

Proposition 16.

The average value of the digit in the jth decimal place is given by

$$10\sum_{k=1}^{10^{j-1}} F\left(\frac{k}{10^{j-1}}\right) - \sum_{k=1}^{10^{j}} F\left(\frac{k}{10^{j}}\right).$$

In particular, for $j = 1$ we have the average value of the first digit equal to

$$10\,(1-\gamma) - \sum_{k=1}^{10} F\left(\frac{k}{10}\right) = 3.736\ldots$$

(Proof) Theorem 1. allows us to translate the words "average value of the digit in the jth decimal place" into the integral

$$\int_0^1 \left(\left[10^j x\right] - 10\left[10^{j-1}x\right]\right) dF\,(x)\,.$$

It is convenient to break this integral into two sums:

$$\sum_{k=0}^{10^j-1}\int_{\frac{k}{10^j}}^{\frac{k+1}{10^j}} k\, dF\,(x) - 10\sum_{k=0}^{10^{j-1}-1}\int_{\frac{k}{10^{j-1}}}^{\frac{k+1}{10^{j-1}}} k\, dF\,(x)$$

Evaluating these sums of integrals we get

$$\sum_{k=1}^{10^j-1} k\left(F\left(\frac{k+1}{10^j}\right) - F\left(\frac{k}{10^j}\right)\right) - 10\sum_{k=1}^{10^{j-1}-1} k\left(F\left(\frac{k+1}{10^{j-1}}\right) - F\left(\frac{k}{10^{j-1}}\right)\right).$$

After what by now has come to be the expected telescoping, we have

$$\left(10^j - 1\right) F(1) - \sum_{k=1}^{10^j-1} F\left(\frac{k}{10^j}\right) -$$

$$10\left(\left(10^{j-1} - 1\right) F(1) - \sum_{k=1}^{10^{j-1}} F\left(\frac{k}{10^{j-1}}\right) + F(1)\right)$$

We only need collect terms to arrive at the desired conclusion:

$$10\sum_{k=1}^{10^{j-1}} F\left(\frac{k}{10^{j-1}}\right) - \sum_{k=1}^{10^{j}} F\left(\frac{k}{10^{j}}\right).$$

The argument in the proof of Proposition 15. gives us

$$\lim_{j\to\infty}\sum_{k=1}^{10^j} F\left(\frac{k}{10^j}\right)=\frac{1}{2}.$$

So the conclusion of Proposition 16. allows us to infer that the average value of the *jth* decimal place converges to $5-\frac{1}{2}=4\frac{1}{2}$. When coupled with Propostion 17. below, this amounts to something like a proof that the average value of the jth decimal place for *the set* of rationals is $4\frac{1}{2}$. Well... We close this section by showing that the tails of the fractional parts are, in the limit, uniformly distributed. Propositions 15., 16.,and 17. show an evaporation of pathology in the fractional parts as we shift into the tails. Of course, Propositions 15. and 16. hold for any base, since we can let 10 represent any positive integer greater than one.

Proposition 17.

In the limit, the tails of the fractional parts are uniformly distributed in [0,1].

(Proof) We must show that the probability of a tail being in the interval (a,b) is $b-a$, in the limit. To accomplish this we fall back on the function ω that we used earlier and Theorem 1. We are faced then with computing the limit as j goes to infinity of

$$\int_0^1 \omega_{[a,b]}\{10^j x\}dF(x).$$

This expression is equivalent to

$$\sum_{k=0}^{10^j-1}\int_{\frac{k}{10^j}+\frac{a}{10^j}}^{\frac{k}{10^j}+\frac{b}{10^j}} dF(x).$$

Moving right along and keeping in mind what we want to show, we evaluate the integral and manipulate the result as follows

$$\sum_{k=0}^{10^j-1}\left(F\left(\frac{k}{10^j}+\frac{b}{10^j}\right)-F\left(\frac{k}{10^j}+\frac{a}{10^j}\right)\right)=$$

$$(b-a)\sum_{k=0}^{10^j-1}\frac{1}{10^j}\frac{F\left(\frac{k}{10^j}+\frac{b}{10^j}\right)-F\left(\frac{k}{10^j}+\frac{a}{10^j}\right)}{\frac{b}{10^j}-\frac{a}{10^j}}.$$

The Mean Value Theorem allows us to replace the difference quotient in the sum with

$$F'(c_{j.k}),$$

where $c_{j.k}$ lies in $(\frac{k}{10^j}+\frac{a}{10^j},\frac{k}{10^j}+\frac{b}{10^j})$, hence in $(\frac{k}{10^j},\frac{k+1}{10^j})$. This transforms our sum into

$$(b-a)\sum_{k=0}^{10^j-1}\frac{1}{10^j}F'(c_{j,k}).$$

Taking the limit as j goes to infinity we get

$$(b-a)\int_0^1 dF(x)=b-a.$$

This completes the proof.

Some Measure-Theoretic Properties

We close with a rather amusing theorem on the measure induced in the unit interval by F. This theorem gives us something like an echo. For a Borel subset A define $\mu(A)$ by

$$\mu(A)=\int_A dF(\theta).$$

Theorem 2. *If*

$$\mu\left(1-\frac{1}{m},1\right)=r_m,$$

then

$$\mu\left(0,r_m\right)=\frac{c(m)}{m},$$

where

$$c(m)\rightarrow 1.06087...=\frac{\pi^2(\pi^2-6)}{36}$$

and

$$1\le c(m)<\frac{3}{2},\ m=1,2,3\ldots$$

This means that

$$\left[\mu\left(0,\mu\left(1-\frac{1}{m},1\right)\right)\right]_m=\left\langle\mu\left(0,\mu\left(1-\frac{1}{m},1\right)\right)\right\rangle_m=\frac{1}{m}.$$

(Proof) We use Proposition 4. to write

$$\mu\left(1-\frac{1}{m},1\right)=\sum_{k=1}^{\infty}\frac{\frac{1}{m}}{(k+1)(k+1-\frac{1}{m})}.$$

It will be convenient to define

$$S(m)=\sum_{k=1}^{\infty}\frac{1}{(k+1)(k+1-\frac{1}{m})}.$$

Then we note that

$$\mu\left(0,\mu\left(1-\frac{1}{m},1\right)\right)=\sum_{k=1}^{\infty}\frac{\frac{1}{m}S(m)}{k(k+\frac{1}{m}S(m))}.$$

We wish to show that

$$\frac{1}{m}\leq\sum_{k=1}^{\infty}\frac{\frac{1}{m}S(m)}{k(k+\frac{1}{m}S(m))}<\frac{3}{2m}.$$

or:

$$\sum_{k=1}^{\infty}\frac{1}{k(k+\frac{1}{m}S(m))}<\frac{3}{2S(m)}\ (*)$$

and

$$\frac{1}{S(m)}\leq\sum_{k=1}^{\infty}\frac{1}{k(k+\frac{1}{m}S(m))}\ (**)$$

The inequalities (*) and (**) hold if $m=1$, since $S(1)=1$. Further, for inequality (*) note that $\frac{1}{m}S(m)$ is a decreasing sequence, so that both expressions are increasing functions of m, with the left side bounded above by $\frac{\pi^2}{6}$, which is smaller than the right hand side for $m>1$.

For inequality (**) we observe that both sides are increasing, but the right side is increasing to $\frac{6}{\pi^2}$, while the left side is increasing to the somewhat smaller value $\frac{1}{\frac{\pi^2}{6}-1}$. The proof is then completed by observing that the inequality holds for the specific integers 2 through 7 inclusive. For $m>7$, the expression on the right is greater than the limiting value $\frac{1}{\frac{\pi^2}{6}-1}$ of $\frac{1}{S(m)}$.

It then follows that

$$c(m)=\frac{\sum_{k=1}^{\infty}\frac{1}{k(k+\frac{1}{m}S(m))}}{\frac{1}{S(m)}}.$$

The above expression converges to

$$\frac{\pi^2(\pi^2-6)}{36}.$$

The method of proof also shows that the theorem holds for sufficiently large real m. It can be shown by a similar, but uglier argument than the one above, that

$$\frac{1}{2m} \leq \sum_{k=1}^{\infty} \frac{\frac{1}{m}S(m)}{k(k+\frac{1}{m}S(m))} < \frac{3}{2m}$$

for all real $m \geq 1$. So that

$$\left\langle \mu\left(0, \mu\left(1-\frac{1}{m}, 1\right)\right)\right\rangle_m = \frac{1}{m}$$

for all real $m \geq 1$.

We conjecture that

$$\left[\mu\left(0, \mu\left(1-\frac{1}{m}, 1\right)\right)\right]_m = \frac{1}{m}$$

for all real $m > \frac{1}{\gamma+\epsilon}$, where epsilon is small.

We also note that Theorem 2 has a natural counterpart. We may substitute

$$\left[\mu\left(1-\mu\left(0, \frac{1}{m}\right), 1\right)\right]_m = \left\langle \mu\left(1-\mu\left(0, \frac{1}{m}\right), 1\right)\right\rangle_m = \frac{1}{m}.$$

If we make the obvious change in the definition of r_m as well, we have the counterpart. The proof is entirely analogous to that of Theorem 2.

Conclusion

Our propositions have involved errors in approximating classical functions and constants. The fact that the iteration of the measure produces an "error" that can be "corrected" to $\frac{1}{m}$ by an application of $[\cdot]_m$ or $\langle\cdot\rangle_m$ seems a fitting end.

We wish to thank the members of our working group *GAMMA* at NSA for their encouragement and willingness to listen. We are indebted to Dan Swearingen, Sandy Perlman, Charlie Toll and Clark Benson. Finally, special thanks are due to Donald J. Newman for his work and leadership in *GAMMA*.

References

1. T. M. Apostol, *An Introduction to Analytic Number Theory*, 1976.
2. G. Arfken, *Mathematical Methods for Physicists*, 1985, pp. 539–555.
3. H. T. Davis, *Tables of the Mathematical Functions, Volume I*, 1963, pp. 277–290.
4. I. S. Gradshteyn and I. M. Ryzhik, *Table of Integrals, Series, and Products*, 1965, pp. 952–956.
5. G.H. Hardy and E.M. Wright, *An Introduction to the Theory of Numbers*, 1995.
6. A. Ya. Khinchin, *Continued Fractions*, 1961.
7. L. M. Milne-Thompson, *The Calculus of Finite Differences*, 1933, pp. 241–270.

Contemporary Mathematics
Volume **275**, 2001

Ultrafilters and Ramsey Theory

Dennis E. Davenport

Abstract. It is well known that given a discrete semigroup $(S, +)$, the operation extends to the Stone-Čech compactification βS of S. Using this extension and the fact that the points of βS are the ultrafilters on S, several powerful results in Ramsey theory can be proved. In this paper we survey some results and we also give an ultrafilter proof of the multidimensional van der Waerdan Theorem.

1. Introduction

Throughout this paper $(S, +)$ will be a discrete semigroup, i.e., $+$ is an associative binary operation and singletons are open in S. There are several definitions one may use for an ultrafilter, we shall use Definition 1.2, a set theoretic definition. See [4] for other ways of defining an ultrafilter.

1.1 Definition. A *filter* $\mathcal{F}$ on S is a subset of $\mathcal{P}(S)$ such that

1. $\emptyset \notin \mathcal{F}$
2. $S \in \mathcal{F}$
3. $\forall A \in \mathcal{F}$, $\forall B \in \mathcal{F}$, $A \cap B \in \mathcal{F}$
4. $\forall A \in \mathcal{F}$, $\forall B \subseteq S$, $A \subseteq B \Rightarrow B \in \mathcal{F}$.

1.2 Definition. A filter p on S is an *ultrafilter* iff $A \cup B \in p \Rightarrow A \in p$ or $B \in p$.

An ultrafilter can also be defined as a maximal filter. Also a filter p is an ultrafilter if and only if $\forall A \subseteq S$, either $A \in p$ or $S \backslash A \in p$. We view points of βS as ultrafilters on S. The following definition describes the extended operation. The proofs of Lemmas 1.4-1.6 can be found in [8].

1.3 Definition. Let $(S, +)$ be a discrete semigroup.

a. $\beta S = \{p : p \text{ is an ultrafilter on } S\}$
b. $\forall A \subseteq S$, $\overline{A} = \{p \in \beta S : A \in p\}$
c. $\forall x \in S$, $e(x) = \{A \subseteq S : x \in A\}$
d. $\forall A \subseteq S$, $\forall x \in S$, $A - x = \{y \in S : y + x \in A\}$
e. $\forall p, q \in \beta S$, $p + q = \{A \subseteq S : \{x \in S : A - x \in p\} \in q\}$. Thus $A \in p + q$ iff $\{x \in S : A - x \in p\} \in q$.

1.4 Lemma. *$\{\overline{A} : A \subseteq S\}$ is a basis for some topology on βS.*

1.5 Lemma. *$\forall x \in S$, $e(x) \in \beta S$. ($e(x)$ is called a* principal *or* fixed *ultrafilter.)*

2000 *Mathematics Subject Classifications.* Primary 05D10, 22A15.

1.6 Lemma. *$(\beta S, +)$ is a semigroup.*

Note that e is a function from S to βS. From now on we will assume that βS has the topology generated by the basis given in Lemma 1.4. With this topology βS is the Stone-Čech compactification of S with embedding e, hence $e[S]$ is dense in βS. We will denote $e(x)$ by x, when this not likely to cause confusion. Also, with the extended operation on βS, e is a semigroup homomorphism, that is, $\forall x, y \in S$, $e(x+y) = e(x) + e(y)$.

The algebraic structure of βS can be used to prove several results in Ramsey Theory. Since βS has both a topological and algebraic structure, a logical question to ask is whether the algebraic operation is continuous. As it turns out, this is false. But for any fixed element p, the function $\lambda_p : \beta S \to \beta S$, given by $\lambda_p(q) = p + q$ is continuous and for each $x \in S$, the function $\rho_x : \beta S \to \beta S$ defined by $\rho_x(p) = p + x$ is continuous.

We now review some semigroup definitions and properties.

1.7 Definition. Let T be a nonempty subset of S.
(a) T is a *right (left) ideal* of S iff $T + S \subseteq T$ $(S + T \subseteq T)$.
(b) T is an *ideal* of S iff T is both a left and right ideal.
(c) An ideal M is the *minimal ideal* of S iff given any ideal J, $J \subseteq M \Rightarrow J = M$.
(d) R is a *minimal right (left) ideal* of S iff R is a right (left) ideal of S and if J is a right (left) ideal of S and $J \subseteq R$, then $J = R$.

1.8 Theorem. *If S has a minimal right (left) ideal, then the minimal ideal M exists and $M = \bigcup\{R : R$ is a mininal right ideal of $S\}$. Also if I is any other ideal of S, then $M \subseteq I$.*

1.9 Definition. Let $(T, +)$ be a semigroup and $\mathcal{T}$ a topology on T. Then $(T, +, \mathcal{T})$ is a *left (right) topological semigroup* iff $\forall x \in T$, λ_x (ρ_x) is continuous, where $\lambda_x : T \to T$ is defined by $\lambda_x(y) = x + y$ and $\rho_x : T \to T$ is defined by $\rho_x(y) = y + x$.

From the above comments we see that βS is a left topological semigroup. The fact that βS is also compact and Hausdorff plays an important role in its algebraic structure. (See Theorems 1.10 and 1.11 below.) By Theorem 1.8 we see that the minimal ideal is the union of all minimal right ideals. And using Zorn's lemma we can show that every compact Hausdorff left topological semigroup has a minimal right ideal. Hence, we get the following theorem.

1.10 Theorem. *Let T be a compact Hausdorff left topological semigroup. Then,*
(a) each right ideal of T contains a minimal right ideal that is closed and
(b) T has a minimal ideal.

The proof of the following theorem can be found in [3]. It is a very elegant application of Zorn's lemma.

1.11 Theorem (Ellis). *Let T be a compact Hausdorff left topological semigroup. Then there exists $t \in T$ such that $t + t = t$ (i.e., T has an idempotent).*

Since βS is a compact, Hausdorff, left topological semigroup, it has both a minimal ideal M and an idempotent. If $S = \mathbb{N}$, the set of natural numbers, then $\beta\mathbb{N}$ has $2^{\mathfrak{c}}$ idempotents.

2. Ramsey Theory

In 1974 Neil Hindman proved the following theorem using combinatorial methods.

2.1 Hindman's Theorem. *Let $r \in \mathbb{N}$. Suppose $\mathbb{N} = \bigcup_{i<r} A_i$. Then there exist $i < r$ and there exists an infinite subset B of $\mathbb{N}$ such that all finite sums of elements from B are contained in A_i.*

Later that same year Glazer, using ideas of Fred Galvin, was able to give an ultrafilter proof of Hindman's theorem. There are four proofs of Hindman's theorem known to the author, but none as nice as Glazer's. Glazer's proof allowed mathematicians to see an unusual connection between two branches of mathematics that seem so different, the theory of topological semigroups and Ramsey theory. Even Hindman prefers the ultrafilter proof. This proof helped generate the study of the algebraic structure of $\beta\mathbb{N}$. It is interesting to note that R. Arens proved that the semigroup operation $+$ extends to $\beta\mathbb{N}$ in 1951.

The first Ramsey theory type theorem known to the author is due to Hilbert.

2.2 Theorem (Hilbert, 1892). *Let $r \in \mathbb{N}$. Suppose $\mathbb{N} = \bigcup_{i<r} A_i$. Then for all $n \in \mathbb{N}$, there exist $i < r$, $a \in \mathbb{N}$, and $a_0, a_1, \dots, a_n$ such that $\left\{a + \sum_{j\in F} a_j : F \subseteq \{0, 1, \dots, n-1\}\right\} \subseteq A_i$.*

In his attempt to prove Fermat's last theorem I. Schur was able to prove the following theorem in 1916.

2.3 Schur's Theorem. *Let $r \in \mathbb{N}$. Suppose $\mathbb{N} = \bigcup_{i<r} A_i$. Then there exist $i < r$ and positive integers x and y such that $\{x, y, x+y\} \subseteq A_i$.*

We can easily find a two cell partition of $\mathbb{N}$ which will guarantee $x \neq y$. Let O be the set of odd positive integers. Let $A_1 = O \cup 4O \cup 16O \cup 64O \cup \dots$ and $A_2 = 2O \cup 8O \cup 32O \cup \dots$. Then $\mathbb{N} = A_1 \cup A_2$, but we cannot find x and i such that $\{x, 2x\} \subseteq A_i$.

Theorems 2.2 and 2.3 can easily be proved using the algebraic structure of $\beta\mathbb{N}$. However, it was not until 1989 when V. Bergelson, H. Furstenberg, N. Hindman, and Y. Katznelson were able to find an algebraic proof of the following theorem.

2.4 van der Waerden's Theorem. *Let $r \in \mathbb{N}$. Suppose $\mathbb{N} = \bigcup_{i<r} A_i$. Then $\forall l \in \mathbb{N}$, there exist natural numbers a, d, and i, such that $\{a + td : t < l\} \subseteq A_i$.*

Hence, there exists $i < r$, such that the set contains arbitrarily long arithmetic progressions. Van der Waerden proved Theorem 2.4 in 1927.

One can easily find a two cell partition of $\mathbb{N}$ to show that arbitrarily long arithmetic progressions cannot be replaced by infinitely long arithmetic progression. Indeed, let $A_1 = \{1, 3, 4, 7, 8, 9, 13, 14, 15, 16, \dots\}$ and $A_2 = \{2, 5, 6, 10, 11, 12, 17, 18, 19, 20, \dots\}$. Then for each j, the set A_j has arbitrarily long gaps. Hence, neither A_1 nor A_2 contains an infinitely long arithmetic progression.

2.5 k-dimensional van der Waerden's Theorem. *Let $k, r \in \mathbb{N}$. Suppose $\mathbb{N}^k = \bigcup_{i<r} A_i$. Then $\forall l \in \mathbb{N}$, $\exists\{a_1, a_2, \dots a_k\} \subseteq \mathbb{N}$, $\exists d \in \mathbb{N}$, $\exists j < r$ such that $\{(a_1 + n_1 d, a_2 + n_2 d, \dots a_k + n_k d) : n_i < l, \forall i\} \subseteq A_i$.*

Note that if we fix $k-1$ coordinates, then we get an arithmetic progression of length l in the coordinate that is allowed to vary. Such a set will be called a k-dimensional arithmetic progression. We shall give an ultrafilter proof of Theorem 2.5. This proof is similar to the proof given by Bergelson, Furstenberg, Hindman, and Katznelson of van der Waerden's Theorem in [2].

3. A Proof of the Multidimensional van der Waerden Theorem

Lemma 3.1 tells us that if we have a partition of an infinite discrete space S, then we can find a partition of βS that relates to the given partition of S in a very natural way. Notice that the algebraic structure is not needed. Recall that if f is a continuous function from S to any compact Hausdorff space X, then there is a continuous extension of f to βS. In this context we view S as a subset of βS, where each $s \in S$ is identified with $e(s)$ the principal ultrafilter. This extension is denoted by f^β.

3.1 Lemma. *Let S be an infinite discrete space. If $\{A_1, A_2, \ldots, A_r\}$ is a partition of S, then $\{c\ell(A_1), \ldots, c\ell(A_r)\}$ is a partition of βS and $c\ell(A_j)$ is open for each $j \leq r$.*

Proof. Let $X = \{1, 2, \ldots, r\}$ with the discrete topology. Define $f : S \to X$ by $f(s) = j \iff s \in A_j$. Since S has the discrete topology, f is continuous. Also X is a compact Hausdorff space, hence f can be extended to βS. Call this extension f^β. For each $j \leq r$, let $B_j = \{p \in \beta S : f^\beta(p) = j\}$. Then $\forall p \in \beta S$, $f^\beta(p) = j$ for some $j \in X$. Thus $\{B_1, B_2, \ldots, B_r\}$ is a partition of βS. Note that $\forall j \in X$, $\{j\}$ is both open and closed, and $B_j = (f^\beta)^{-1}[\{j\}]$. Since f^β is continuous, B_j is both open and closed. We claim that $\forall j \leq r$, $B_j = c\ell_{\beta S}(A_j)$. Indeed, $A_j \subseteq B_j$ so $c\ell(A_j) \subseteq B_j$. Now let $p \in B_j$ and $A \in p$. Since S is dense in βS and $\overline{A} \cap B_j$ is a nonempty open set, pick $s \in S \cap (\overline{A} \cap B_j)$. Thus $s \in B_j$ so $f^\beta(s) = j$. This implies $f(s) = j$. Therefore $s \in A_j$, so $s \in \overline{A} \cap A_j$. Hence $p \in c\ell_{\beta S}(A_j)$. Therefore $B_j = c\ell_{\beta S}(A_j)$. □

For the remainder of this paper we fix a natural number l. Our goal is to show that for a given finite partition of $\mathbb{N}^k$, we can find a member of the partition such that this member contains a k-dimensional arithmetic progression of length l. Then by the Pigeon-Hole principle we can find a member of the partition that contains arbitrarily long k-dimensional arithmetic progressions. Also, we will order $\mathbb{N}^k$ lexicographically; that is, if $a, b \in \mathbb{N}^k$, where $a = (a_1, \ldots, a_k)$ and $b = (b_1, \ldots, b_k)$, then $a \leq b \iff a_i = b_i$ for all i or $a_n < b_n$ at the first place where they differ.

3.2 Definition.
(a) Let $X = [\beta(\mathbb{N}^k)]^{l^k}$ with the product topology and coordinatewise operations.
(b) For each $a \in \mathbb{N}$ and $d \in \omega$, let $A(a, d) = \{a + jd : 0 \leq j < l\}$.
(c) Let $\widehat{E} = \{(b_1, \ldots, b_{l^k}) : b_i = (c_{i_1}, \ldots, c_{i_k})$, where $c_{i_j} \in A(a_j, d)$, for $j = 1, 2, \ldots, k$, $\{a_1, a_2, \ldots, a_k\} \subseteq \mathbb{N}$, $d \in \omega$, and $b_m < b_{m+1}, \forall m\}$.
(d) Let $\widehat{I} = \{(b_1, \ldots, b_{l^k}) : b_i = (c_{i_1}, \ldots, c_{i_k})$, where $c_{i_j} \in A(a_j, d)$, for $j = 1, 2, \ldots, k$, $\{a_1, a_2, \ldots, a_k\} \subseteq \mathbb{N}$, $d \in \mathbb{N}$, and $b_m < b_{m+1}, \forall m\}$.
(e) Let $E = c\ell\widehat{E}$ and $I = c\ell\widehat{I}$.

One can easily prove that X is a compact Hausdorff left topological semigroup, hence X has a minimal ideal and X has idempotents. Also, if $x = (x_1, \ldots, x_{l^k}) \in \mathbb{N}^{l^k}$, then ρ_x is continuous.

3.3 Lemma. *Let $x \in \widehat{E}$.*
(i) If $y \in \widehat{E}$, then $x + y \in \widehat{E}$;
(ii) If $y \in \widehat{I}$, then $x + y, y + x \in \widehat{I}$.

Proof. We will prove (i) and leave the proof of (ii) to the reader. Since x and y are elements of $\widehat{E}$, there exists $d(x), d(y) \in \omega$ and there exists $a(x)_1, \ldots, a(x)_k, a(y)_1, \ldots, a(y)_k \in \mathbb{N}$ such that $x = (x_1, \ldots, x_{l^k})$ and $y = (y_1, \ldots, y_{l^k})$, where $x_i = (b_{i_1}, \ldots, b_{i_k})$ and $y_i = (c_{i_1}, \ldots, c_{i_k})$ with $b_{i_j} \in A(a(x)_j, d(x))$ and $c_{i_j} \in A(a(y)_j, d(y))$, and $x_i < x_{i+1}$ and $y_i < y_{i+1}$, for $i = 1, \ldots, l^k - 1$. Now $x + y = (x_1 + y_1, \ldots, x_{l^k} + y_{l^k})$ and since $x_i < x_{i+1}$ and $y_i < y_{i+1}$, for $i = 1, \ldots, l^k - 1$, $x_i + y_i < x_{i+1} + y_{i+1}$, for $i = 1, \ldots, l^k - 1$. Also $x_i + y_i = (b_{i_1} + c_{i_1}, \ldots, b_{i_k} + c_{i_k})$ where for each $j = 1, \ldots, k$, $b_{i_j} = a(x)_j + nd(x)$ and $c_{i_j} = a(y)_j + md(y)$, with $0 \leq n < l$ and $0 \leq m < l$. Since $x_i < x_{i+1}$ and $y_i < y_{i+1}$, for $i = 1, \ldots, l^k - 1$, we must have $m = n$. Thus $c_{i_j} + b_{i_j} \in A(a(x)_j + a(y)_j, d(x) + d(y))$. Hence $x + y \in \widehat{E}$. □

3.4 Lemma. *E is a subsemigroup of X and I is an ideal of E.*

Proof. Let $p, q \in E$. We claim that $p + q \in E$. Let U be an open neighborhood of $p + q$. Since X is a left topological semigroup, λ_p is continuous at q. Hence, we can pick a neighborhood V of q such that $p + V \subseteq U$. Since $q \in E$, we can pick $a \in \widehat{E}$ such that $a \in V$. Thus $p + a \in U$. Since ρ_a is continuous, we can find a neighborhood W of p such that $W + a \subseteq U$. Now $p \in E$, so $W \cap \widehat{E} \neq \emptyset$. Thus, we can pick $b \in W \cap \widehat{E}$. Hence $b + a \in U$. By Lemma 3.3, we know $b + a \in \widehat{E}$. Also, if $p \in I$ and $q \in E$, a similar argument shows both $p + q$ and $q + p$ are elements of I.□

We are now in the position to prove our main theorem about the algebraic structure of $(\beta\mathbb{N}^k)^{l^k}$. Recall that $\mathbb{N}^k$ is a discrete semigroup, where the semigroup operation is defined coordinatewise. Hence $\beta(\mathbb{N}^k)$ is a compact Hausdorff left topological semigroup. Thus the minimal ideal of $\beta(\mathbb{N}^k)$ exists, we shall denote the minimal ideal by $M(\beta(\mathbb{N}^k))$.

3.5 Theorem. *Let $p \in M(\beta(\mathbb{N}^k))$ and let $\overline{p} = (p, p, \ldots, p)$, with l^k coordinates. Then $\overline{p} \in I$.*

Proof. Let $U_1 \times U_2 \times \ldots \times U_{l^k}$ be a basic neighborhood of $\overline{p}$. Thus $\bigcap_{i=1}^{l^k} U_i \in p$. Now $\mathbb{N}^k$ is dense in $\beta(\mathbb{N}^k)$, so we can pick $(a_1, \ldots, a_k) \in \mathbb{N}^k$ such that $(a_1, \ldots, a_k) \in \bigcap_{i=1}^{l^k} U_i$. Note that $((a_1, \ldots, a_k), (a_1, \ldots, a_k), \ldots, (a_1, \ldots, a_k)) \in \widehat{E}$. Hence

$$((a_1, \ldots, a_k), (a_1, \ldots, a_k), \ldots, (a_1, \ldots, a_k)) \in \widehat{E} \cap \prod_{i=1}^{l^k} U_i \,.$$

Thus $\overline{p} \in c\ell(\widehat{E}) = E$.

Now $p \in M(\beta(\mathbb{N}^k))$, so by Theorem 1.7, we can pick a minimal right ideal R of $\beta(\mathbb{N}^k)$ such that $p \in R$. Clearly $\overline{p} + E$ is a right ideal of E and E is a compact Hausdorff left topological semigroup, hence by Theorem 1.10(a), we can find a closed minimal right ideal T of E such that $T \subseteq \overline{p} + E$. Since T is a closed subset of the compact space E, T is also compact. Hence T is a compact Hausdorff left topological semigroup. Pick by Theorem 1.11, $\overline{t} = (t_1, \ldots, t_{l^k}) \in T$ such that $\overline{t}$ is an idempotent. Note that for $i = 1, \ldots, l^k$, t_i is an idempotent of $\beta(\mathbb{N}^k)$. So $t \in \overline{p} + E$. Thus we can pick $\overline{s} = (s_1, \ldots, s_{l^k}) \in E$ such that $\overline{t} = \overline{p} + \overline{s}$.

We claim that $\overline{t} + \overline{p} = \overline{p}$. Indeed, let $i \in \{1, 2, \ldots, l^k\}$. Then $t_i = p + s_i$. Since $p \in R$ ar : R is a right ideal, $t_i = p + s_i \in R$. Thus $t_i + \beta(\mathbb{N}^k) \subseteq R + \beta(\mathbb{N}^k) \subseteq R$. Since R is a minimal right ideal and $t_i + \beta(\mathbb{N}^k)$ is a right ideal, $t_i + \beta(\mathbb{N}^k) = R$. Since $p \in R$, we can pick $q \in \beta(\mathbb{N}^k)$ such that $t_i + q_i = p$. Thus $t_i + p = t_i + t_i + q_i = t_i + q_i = p$. Therefore $\overline{t} + \overline{p} = \overline{p}$.

Now $\overline{t} \in T$ and T is a minimal right ideal of E, hence $\overline{p} = \overline{t} + \overline{p} \in T$. Also by Theorem 1.8, $T \subseteq M(E) \subseteq I$. Thus $\overline{p} \in I$. □

We are now in the position to prove Theorem 2.5 as a corollary to Theorem 3.5.

2.5 k-dimensional van der Waerden's Theorem. *Let $k, r \in \mathbb{N}$. Suppose $\mathbb{N}^k = \bigcup_{i<r} A_i$. Then $\forall l \in \mathbb{N}$, $\exists\{a_1, a_2, \ldots a_k\} \subseteq \mathbb{N}$, $\exists d \in \mathbb{N}$, $\exists j < r$ such that $\{(a_1 + n_1 d, a_2 + n_2 d, \ldots a_k + n_k d) : n_i < l, \forall i\} \subseteq A_i$.*

Proof. Since $\beta(\mathbb{N}^k)$ is a compact Hausdorff left topological semigroup, by Theorem 1.10, $M(\beta(\mathbb{N}^k)) \neq \emptyset$. Pick $p \in M(\beta(\mathbb{N}^k))$ and let $\overline{p} = (p, p, \ldots, p)$, (with l^k coordinates). By Lemma 3.1, choose $i < r$ such that $p \in c\ell_{\beta(\mathbb{N}^k)}(A_i)$ and let $U = c\ell_{\beta(\mathbb{N}^k)}(A_i)$. Then $U \times U \times \ldots \times U$ is a neighborhood of $\overline{p}$. By Theorem 3.5, $\overline{p} \in c\ell_X(\widehat{I}) = I$. Hence $U \times U \times \ldots \times U \cap \widehat{I} \neq \emptyset$. Pick $a_1, \ldots, a_k \in \mathbb{N}$ and $d \in \mathbb{N}$ such that $(b_1, \ldots, b_{l^k}) \in U \times U \times \ldots \times U$, where $b_i = (c_{i_1}, \ldots, c_{i_k})$, where $c_{i_j} \in A(a_j, d)$, for $j = 1, 2, \ldots, k$, and $b_m < b_{m+1}$, for $m = 1, \ldots, l^k$. Thus $b_j \in U \cap \mathbb{N}^k = A_i$. □

In [5], H. Furstenberg and Y. Katznelson used similar methods to prove the Hales-Jewett Theorem. They started with the free semigroup on l letters instead of $\mathbb{N}^k$.

References

1. R. Arens, *The adjoint of a bilinear operation*, Proc. Amer. Math. Soc. **2** (1951), 839-848.
2. V. Bergelson, H. Furstenberg, N. Hindman, and Y. Katznelson, *An algebraic proof of van der Waerden's theorem*, Enseign. Math. **35** (1989), 209-215.
3. J. Berglund, H. Junghenn, P. Milnes, *Analysis on semigroups*, Wiley, N.Y., 1989.
4. A. Blass, *Ultrafilters: where topological dynamics = algebra = combinatorics*, Topology Proc. **18** (1993), 33-56.
5. H. Furstenberg and Y. Katznelson, *Idempotents in compact semigroups and Ramsey theory*, Irsrael J. Math. **68** (1989), 257-270.
6. R. Graham, B. Rothschild, and J. Spencer, *Ramsey Theory*, Wiley, N.Y., 1990.
7. N. Hindman, *Finite sums from sequences within cells of a partition of* $\mathbb{N}$, J. Comb. Theorey (Series A) **17** (1974), 1-11.
8. N. Hindman, *Ultrafilters and combinatorial number theory*, in Number Theory Carbondale 1979 (M. Nathanson, ed.), Lecture Notes in Math. **751** (1979), 119-184.

Contemporary Mathematics
Volume **275**, 2001

Artin's Conjecture and Elliptic Curves

Edray Herber Goins

ABSTRACT. Artin conjectured that certain Galois representations should give rise to entire L-series. We give some history on the conjecture and motivation of why it should be true by discussing the one-dimensional case. The first known example to verify the conjecture in the icosahedral case did not surface until Buhler's work in 1977. We explain how this icosahedral representation is attached to a modular elliptic curve isogenous to its Galois conjugates, and then explain how it is associated to a cusp form of weight 5 with level prime to 5.

1. Introduction

In 1917, Erich Hecke [**10**] proved a series of results about certain characters which are now commonly referred to as Hecke characters; one corollary states that one-dimensional complex Galois representations give rise to entire L-series. He revealed, through a series of lectures [**9**] at Princeton's Institute for Advanced Study in the years that followed, the relationship between such representations as generalizations of Dirichlet characters and modular forms as the eigenfunctions of a set of commuting self-adjoint operators. Many mathematicians were inspired by his ground-breaking insight and novel proof of the analytic continuation of the L-series.

In the 1930's, Emil Artin [**1**] conjectured that a generalization of such a result should be true; that is, irreducible complex projective representations of finite Galois groups should also give rise to entire L-series. He came to this conclusion after proving himself that both 3-dimensional and 4-dimensional representations of the simple group of order 60, the alternating group on five letters, might give rise to L-series with singularities. It is known, due to the insight of Robert Langlands [**16**] in the 1970's relating Hecke characters with Representation Theory, that in order to prove the conjecture it suffices to prove that such representations are associated to cusp forms. This conjecture has been the motivation for much study in both Algebraic and Analytic Number Theory ever since.

In this paper, we present an elementary approach to Artin's Conjecture by considering the problem over $\mathbb{Q}$. We consider Dirichlet's theorem which preceeded Hecke's results, and sketch a proof by introducing theta series. We then introduce Langland's program to exhibit cusp forms. We conclude by studying a specific example which is associated to an elliptic curve. We assume in the final sections that the reader is somewhat familiar with the basic properties of elliptic curves.

---

[1]This research was sponsored by a joint fellowship from the National Physical Science Consortium (NSPC) and the National Security Agency (NSA), as well as a grant from the NSF, number DMS 97-29992.

1991 *Mathematics Subject Classification.* Primary 11F80; Secondary 11F03, 11F66.

2. One-Dimensional Representations

We begin with some classical definitions and theorems. We are motivated by the expositions in [**11**] and [**4**].

2.1. Reimann Zeta Function and Dirichlet L-Series. Let N be a fixed positive integer, and $\chi : (\mathbb{Z}/N\mathbb{Z})^{\times} \to \mathbb{C}^{\times}$ be a group homomorphism. We extend $\chi : \mathbb{Z} \to \mathbb{C}$ to the entire ring of integers by defining 1) $\chi(n) = \chi(n \mod N)$ on the residue class modulo N; and 2) $\chi(n) = 0$ if n and N have a factor in common. One easily checks that this extended definition still yields a multiplicative map i.e. $\chi(n_1\, n_2) = \chi(n_1)\,\chi(n_2)$.

Fix a complex number $s \in \mathbb{C}$ and associate the *L-series* to χ as the sum

$$L(\chi, s) = \chi(1) + \frac{\chi(2)}{2^s} + \frac{\chi(3)}{3^s} + \cdots = \sum_{n=1}^{\infty} \frac{\chi(n)}{n^s} \tag{2.1}$$

Quite naturaly, two questions arise:

For which region is this series a well-defined function?

Can that function be continued analytically to the entire complex plane?

This series is reminiscent of the *Reimann zeta function*, defined as

$$\zeta(s) = 1 + \frac{1}{2^s} + \frac{1}{3^s} + \cdots = \sum_{n=1}^{\infty} \frac{1}{n^s} \tag{2.2}$$

We may express this sum in a slightly different fashion. It is easy to check that[1]

$$\int_0^{\infty} e^{-ny}\, y^{s-1}\, dy = \frac{(s-1)!}{n^s} \tag{2.3}$$

so that the sum becomes

$$\zeta(s) = \sum_{n=1}^{\infty} \frac{1}{n^s} = \sum_{n=1}^{\infty} \frac{1}{(s-1)!} \int_0^{\infty} e^{-ny}\, y^{s-1}\, dy = \frac{1}{(s-1)!} \int_0^{\infty} \frac{y^{s-1}}{e^y - 1}\, dy \tag{2.4}$$

The function e^y grows faster that any power of y, so the integrand converges for all complex $s-1$ with positive real part. However, we must be careful when $y = 0$; the denominator of the integral vanishes, so we need the numerator to vanish as well. This is not the case if $s = 1$. Hence, we find the

THEOREM 2.1 (Dirichlet [**11**]). *$L(\chi, s)$ is a convergent series if $Re(s) > 1$. If χ is not the trivial character $\chi_0 = 1$ (i.e. $N \neq 1$) then $L(\chi, s)$ has analytic continuation everywhere. On the other hand, $L(\chi_0, s) = \zeta(s)$ has a pole at $s = 1$. In either case, the L-series has the product expansion*

$$L(\chi, s) = \prod_{primes\ p} \left(1 - \frac{\chi(p)}{p^s}\right)^{-1} \tag{2.5}$$

---

[1] In the literature, it is standard to define

$$\Gamma(s) = \int_0^{\infty} e^{-y}\, y^{s-1}\, dy = (s-1)!, \qquad \text{Re}(s) \geq 1$$

as the Generalized Factorial or Gamma function.

PROOF. We present a sketch of proof when $N = 5$, although the method works in general. Consider the character

$$\left(\frac{5}{n}\right) = \begin{cases} +1 & \text{if } n \equiv 1,\, 4 \mod 5; \\ -1 & \text{if } n \equiv 2,\, 3 \mod 5; \\ 0 & \text{if } n \text{ is divisible by } 5. \end{cases} \tag{2.6}$$

The L-series in this case may be expressed as the integral

$$L\left(\left(\frac{5}{*}\right), s\right) = \frac{1}{(s-1)!} \int_0^\infty \left[\sum_{n=1}^\infty \left(\frac{5}{n}\right) e^{-ny}\right] y^{s-1}\, dy \tag{2.7}$$

The infinite sum inside the integrand can be evaulated by considering different cases: When n is divisible by 5, the character vanishes. Otherwise write $n = 5\,k+1, \dots, 5\,k+4$ in the other cases so that the sum becomes

$$\begin{aligned} \sum_{n=1}^\infty \left(\frac{5}{n}\right) e^{-ny} &= \sum_{k=0}^\infty \left[+e^{-(5k+1)y} - e^{-(5k+2)y} - e^{-(5k+3)y} + e^{-(5k+4)y}\right] \\ &= \frac{e^{-y} - e^{-2y} - e^{-3y} + e^{-4y}}{1 - e^{-5y}} = \frac{e^{3y} - e^{y}}{1 + e^{y} + e^{2y} + e^{3y} + e^{4y}} \end{aligned} \tag{2.8}$$

which gives the L-series as

$$L\left(\left(\frac{5}{*}\right), s\right) = \frac{1}{(s-1)!} \int_0^\infty \frac{e^{3y} - e^{y}}{1 + e^{y} + e^{2y} + e^{3y} + e^{4y}}\, y^{s-1}\, dy \tag{2.9}$$

Hence the integrand is analytic at $y = 0$, so the L-series does not have a pole at $s = 1$. □

2.2. Galois Representations. Fix $q(x)$ as an irreducible polynomial of degree d with leading coefficient 1 and integer coefficients, and set $K/\mathbb{Q}$ as its splitting field. The group of permutations of the roots $\mathrm{Gal}(K/\mathbb{Q})$ has a canonical representation as $d \times d$ matrices. To see why, write the d roots q_k of $q(x)$ as d-dimensional unit vectors:

$$q_1 = \begin{pmatrix} 1 \\ 0 \\ \vdots \\ 0 \end{pmatrix}; \quad q_2 = \begin{pmatrix} 0 \\ 1 \\ \vdots \\ 0 \end{pmatrix}; \quad \dots; \quad q_d = \begin{pmatrix} 0 \\ 0 \\ \vdots \\ 1 \end{pmatrix}. \tag{2.10}$$

Any permutation σ on these roots may be representated as a $d \times d$ matrix $\varrho(\sigma)$. One easily checks that ϱ is a multiplicative map i.e. $\varrho(\sigma_1\, \sigma_2) = \varrho(\sigma_1)\, \varrho(\sigma_2)$. As an example, consider $x^2 + x - 1$. Then $K = \mathbb{Q}(\sqrt{5})$, and the only permutation of interest is

$$\sigma : \frac{-1+\sqrt{5}}{2} \mapsto \frac{-1-\sqrt{5}}{2} \quad \Longrightarrow \quad \varrho(\sigma) = \begin{pmatrix} & 1 \\ 1 & \end{pmatrix}. \tag{2.11}$$

Denote $G_{\mathbb{Q}}$, the *absolute Galois group*, as the union of each of the $\mathrm{Gal}(K/\mathbb{Q})$ for such polynomials $q(x)$ with integer coefficients. This larger group still permutes the roots of a specific polynomial $q(x)$; we consider it because it is a universal object independent of $q(x)$. We view the permutation representation as a group homomorphism $\varrho : G_{\mathbb{Q}} \to GL_d(\mathbb{C})$.

The finite collection of matrices $\{\varrho(\sigma) | \sigma \in G_{\mathbb{Q}}\}$ acts on the d-dimensional complex vector space $\mathbb{C}^d$, so we are concerned with *lines* which are invariant under the action of all of the $\varrho(\sigma)$. That is, if we attempt to simultantously diagonalize all of the $\varrho(\sigma)$ then we want to consider one-dimensional invariant subspaces. For example, for $x^2 + x - 1$ we may instead choose the basis

$$q_1{}' = \frac{1}{\sqrt{2}} \begin{pmatrix} 1 \\ 1 \end{pmatrix} \quad q_2{}' = \frac{1}{\sqrt{2}} \begin{pmatrix} 1 \\ -1 \end{pmatrix} \qquad \Longrightarrow \qquad \varrho'(\sigma) = \begin{pmatrix} 1 & \\ & -1 \end{pmatrix}; \tag{2.12}$$

so that we have the more intuitive representation $\rho(\sigma) = -1$ defined on the eigenvalues. In general, we do not wish to consider the $d \times d$ matrix representation ϱ, but rather the scalar $\rho : G_{\mathbb{Q}} \to \mathbb{C}^\times$.

In order to define an L-series, we use the product expansion found in 2.1 above. To this end, choose a prime number p and factor $q(x)$ modulo p, say in the form

$$q(x) \equiv \left(x^{f_1} + \dots\right)^{e_1} \left(x^{f_2} + \dots\right)^{e_2} \dots \left(x^{f_r} + \dots\right)^{e_r} \mod p \tag{2.13}$$

If each $e_j = 1$, we say p is *unramified* (and *ramified* otherwise). In this case, there is a universal automorphism $\mathrm{Frob}_p \in G_{\mathbb{Q}}$ which yields surjective maps $G_{\mathbb{Q}} \to \mathbb{Z}/f_j\,\mathbb{Z}$ for $j = 1, \dots, r$. This automorphism is canonically defined by the congruence

$$\mathrm{Frob}_p\,(q_k) \equiv q_k{}^p \mod p \qquad \text{on the roots } q_k \text{ of } q(x). \tag{2.14}$$

Unfortunately, when p is ramified there is not a canonical choice of Frobenius element because there are repeated roots. We will denote Σ as a finite set containing the ramified primes.

If ρ is a map $G_{\mathbb{Q}} \to \mathbb{C}^\times$, the Frobenius element induces a map $G_{\mathbb{Q}} \to \mathbb{Z}/f_j\,\mathbb{Z}$, and Dirichlet characters are maps $\mathbb{Z}/f_j\,\mathbb{Z} \to \mathbb{C}$, it should follow that Dirichlet characters χ are clcsely related to such maps ρ. This is indeed the case.

THEOREM 2.2 (Artin Reciprocity [**11**]). *Fix $q(x)$ and $\rho : G_{\mathbb{Q}} \to \mathbb{C}^\times$ be as described above. Define $\chi_\rho : \mathbb{Z} \to \mathbb{C}$ on primes by the identification*

$$\chi_\rho(p) = \begin{cases} \rho(\mathit{Frob}_p) & \text{when unramified,} \\ 0 & \text{when ramified;} \end{cases} \tag{2.15}$$

and extend χ_ρ to all of $\mathbb{Z}$ by multiplication. Then there exists a positive integer $N = N(\rho)$, called the conductor of ρ, divisible only by the primes which ramify such that χ_ρ is a Dirichlet character modulo N.

As an example, $x^2 + x - 1$ factors modulo the first few primes as

$$\begin{aligned} x^2 + x - 1 &\equiv x^2 + x + 1 && \mod 2 \\ &\equiv x^2 + x + 2 && \mod 3 \\ &\equiv (x+3)^2 && \mod 5 \\ &\equiv x^2 + x + 6 && \mod 7 \\ &\equiv (x+4)(x+8) && \mod 11 \end{aligned} \tag{2.16}$$

Then Frob_2, Frob_3, and Frob_7 are each nontrivial automorphisms, while Frob_{11} is the identity. The only ramified prime is 5 because of the repeated roots. We map

$$\rho(\mathrm{Frob}_2) = \rho(\mathrm{Frob}_3) = \rho(\mathrm{Frob}_7) = -1; \qquad \rho(\mathrm{Frob}_{11}) = +1. \tag{2.17}$$

The asociated Dirichlet character $\chi_\rho = \left(\frac{5}{*}\right)$ is just the character modulo $N = 5$ we considered above.

2.3. Artin L-Series: 1-Dimensional Case. We are now in a position to define and study the L-series associated to Galois representations.

COROLLARY 2.3 (Artin, Dirichet). *Given a map $\rho : G_\mathbb{Q} \to \mathbb{C}^\times$ as defined above with Σ a finite set containing the ramified primes, define the Artin L-series as*

$$L_\Sigma(\rho,\, s) = \sum_{n=1}^{\infty} \frac{\chi_\rho(n)}{n^s} = \prod_{p \notin \Sigma} \left(1 - \frac{\rho(Frob_p)}{p^s}\right)^{-1} \tag{2.18}$$

Then $L(\rho,\, s)$ converges if $Re(s) > 1$. If ρ is not the trivial map $\rho_0 = 1$ then $L_\Sigma(\rho,\, s)$ has analytic continuation everywhere. On the other hand,

$$L_\Sigma(\rho_0,\, s) = \zeta(s) \cdot \prod_{p \in \Sigma} (1 - p^s) \tag{2.19}$$

has a pole at $s = 1$.

PROOF. We sketch the proof. First, we express the L-series in terms of the integral of a function which dies exponentially fast. This will guarantee that the L-series converges for some right-half plane. For $\tau = x + i\,y$ in the upper-half plane (i.e. $y > 0$) define

$$\theta_\rho(\tau) = \sum_{n=1}^{\infty} \chi_\rho(n)\, n^\epsilon\, e^{\pi i\, n^2\, \tau} \qquad \text{where} \qquad \epsilon = \begin{cases} 0 & \text{if } \chi_\rho(-1) = +1, \\ 1 & \text{if } \chi_\rho(-1) = -1; \end{cases} \tag{2.20}$$

so that the L-series may be expressed as the integral

$$L(\rho,\, s) = \frac{\pi^{\frac{s+\epsilon}{2}}}{\left(\frac{s+\epsilon-2}{2}\right)!} \int_0^\infty \theta_\rho(iy)\, y^{\frac{s+\epsilon-2}{2}}\, dy \tag{2.21}$$

The integrand may have a pole at $y = 0$, so we perform our second trick. Break the integral up into two regions $0 < y < 1$ and $1 < y$, and then use the functional equation

$$\theta_{\overline{\rho}}\left(-\frac{1}{N^2\,\tau}\right) = w(\rho)\, N^{\epsilon+\frac{1}{2}}\, \tau^{\epsilon+\frac{1}{2}}\, \theta_\rho(\tau) \tag{2.22}$$

— where $w(\rho)$ is a complex number of absolute value 1, $N = N(\rho)$ is the conductor, and $\overline{\rho}$ is the complex conjugate — to express the integral solely in terms of values $1/N < y$:

(2.23)
$$\int_0^\infty \theta_\rho(iy)\, y^{\frac{s+\epsilon-2}{2}}\, dy = \int_{1/N}^\infty \left[\theta_\rho(iy)\, y^{\frac{s+\epsilon-2}{2}} + w(\overline{\rho})\, N^{-s+\frac{1}{2}}\, \theta_{\overline{\rho}}(iy)\, y^{\frac{-s+\epsilon-1}{2}}\right] dy$$

Hence the integral converges for all complex s. □

3. Artin's Conjecture

3.1. Artin L-Series: General Case. Fix $q(x)$ as an irreducible polynomial of degree d with leading coefficient 1 and integer coefficients, let Σ be a finite set containing the primes which ramify, and set $G_\mathbb{Q}$ as the absolute Galois group as before. We found that there is a canonical permutation $\varrho : G_\mathbb{Q} \to GL_d(\mathbb{C})$ induced by the action on the roots. Consider the ring

$$V_\varrho = \left\{ \sum_{\sigma \in G_\mathbb{Q}} \lambda_\sigma\, \varrho(\sigma) \in GL_d(\mathbb{C}) \,\middle|\, \lambda_\sigma \in \mathbb{C} \right\} \tag{3.1}$$

generated by the linear combinations of matrices in the image of ϱ. We view this as a complex vector space, which is acted upon by the linear transformations $\varrho(\sigma)$ quite naturally by matrix multiplication. As with any complex vector space, we may decompose it into invariant subspaces. Before, we considered only one-dimensional spaces, but now we generalize to an arbitrary invariant irreducible subspace $V \subseteq V_\varrho$. We restrict ϱ such that the action is faithful on this subspace. That is,

$$\rho = \varrho|_V : \quad G_\mathbb{Q} \to GL(V) \qquad \text{is irreducible.} \tag{3.2}$$

Define the L-series associated to ρ as the product

$$L_\Sigma(\rho,\, s) = \prod_{p \notin \Sigma} \det\left(1 - \frac{\rho(\mathrm{Frob}_p)}{p^s}\right)^{-1} = \sum_{n=1}^\infty \frac{a_\rho(n)}{n^s} \tag{3.3}$$

As before, there is a corresponding $N = N(\rho)$, called the *conductor*, which is divisible by the primes $p \in \Sigma$. By considering the determinant, we find a Dirichlet character ϵ_ρ, called the *nebentype*, which is associated to the one-dimensional Galois representation $\det \circ \rho$. The coefficients $a_\rho(n)$ are closely related to the Frobenius element:

$$a_\rho(p) = \begin{cases} \mathrm{tr}\, \rho(\mathrm{Frob}_p) & \text{when unramified,} \\ 0 & \text{when ramified.} \end{cases} \tag{3.4}$$

Quite naturally, two questions arise once again:

For which complex numbers s is this series a well-defined function?
Can that function be continued analytically to the entire complex plane?

The first question has an answer which is consistent with the theme so far.

PROPOSITION 3.1 (Artin [**1**]). *$L_\Sigma(\rho,\, s)$ converges if $Re(s) > 1$. The L-series associated to the trivial map $L_\Sigma(\rho_0,\, s)$ has a pole at $s = 1$.*

However, the second question remains an open problem.

CONJECTURE 3.2 (Artin). *If ρ is irreducible, not trivial, and is unramified outside a finite set of primes Σ, then $L_\Sigma(\rho, s)$ has analytic continuation everywhere.*

The case of one-dimensional Galois representations (i.e. $V \simeq \mathbb{C}$) was proved in full generality with the advent of Class Field Theory. Indeed, any one-dimensional Galois representation must necessarily be abelian, so that by Artin Reciprocity the representation can be associated with a character defined on the idele group. When working over $\mathbb{Q}$, this amounts to saying that every one-dimensional representation may be associated with a Dirichlet character.

Many mathematicians, inspired by this result, began work on the irreducible two-dimensional representations (i.e. $V \simeq \mathbb{C}^2$). Felix Klein [**12**] had showed that the only finite projective images in the complex general linear group correspond to the Platonic Solids; that is, they are the rotations of the regular polygons and regular polyhedra. This is because we have the injective map

$$SO_3(\mathbb{R}) = \left\{ \gamma \in \mathrm{Mat}_3(\mathbb{R}) \middle| \det \gamma = 1, \ \gamma^t = \gamma^{-1} \right\} \to PGL_2(\mathbb{C}) \tag{3.5}$$

which relates rotations of three-dimensional symmetric objects with 2×2 matrices modulo scalars. Hence, it suffices to consider irreducible two-dimensional projective representations with these images in order to prove Artin's Conjecture in this case.

Most of these cases of the conjecture have been answered in the affirmative. Irreducible cyclic and dihedral representations (that is, representations whose image in $PGL_2(\mathbb{C})$ is isomorphic to Z_n or D_n, respectively) may be interpreted as representations induced from abelian ones, so that the proof of analytic continuation may be reduced to one using Dirichlet characters. Irreducible tetrahedral and some octahedral representations (i.e. projective image isomorphic to A_4 or S_4, respectively) were proved to give entire L-series due to work by Robert Langlands [**15**] in the 1970's on base change for $GL(2)$. The remaining cases for irreducible octahedral representations were proved shortly thereafter by Jerrold Tunnell [**19**]. Such methods worked because they exploited the existence proper nontrivial normal subgroups. Unfortunately, the simple group of order 60 has none, so it is still not known whether the irreducible icosahedral representations (i.e. projective image isomorphic to A_5) have analytic continuation. The first known example to verify Artin's conjecture in this case did not surface until Joe Buhler's work [**3**] in 1977. It is this example we wish to consider in detail.

For general finite dimensional representations $\rho : G_\mathbb{Q} \to GL(V)$, not much is known. It is easy to show that the L-series is analytic in a right-half of the complex plane. In 1947, Richard Brauer [**2**] proved that the characters associated to representations of finite groups are a finite linear combination of one-dimensional characters, and so the corresponding L-series have *meromorphic* continuation; that is, the functions have at worst poles at a finite number of places. Brauer's proof does not guarantee that the integral coefficients of such a linear combination are positive; it can be shown that in many cases the coefficients are negative so that the proof of continuation to the entire complex plane may be reduced to showing that the poles of the L-series are cancelled by the zeroes.

3.2. Maass Forms and the Langlands Program. Robert Langlands completed a circle of ideas which related L-series, complex representations, and automorphic representations. While the deep significance of these ideas is far beyond the scope of this paper, we will content ourselves with a simplified consequence.

THEOREM 3.3 (Langlands [**15**]). *In order to prove Artin's Conjecture for two-dimensional representations ρ unramified outside a finite set of primes, it suffices to prove that*

$$f_\rho(\tau) = \sum_{n=0}^{\infty} a_\rho(n)\, y^{\frac{k}{2}}\, e^{2\pi i\, n\, \tau} \qquad (\tau = x + iy) \tag{3.6}$$

is a Maass cusp form of weight $k = 1$, level $N = N(\rho)$, and nebentype $\epsilon_\rho = \det \circ \rho$.

The idea is to express the L-series as the integral of a function which dies exponentially fast. Indeed, we have the relation

$$L_\Sigma(\rho, s) = \frac{2^s\, \pi^s}{(s-1)!} \int_0^\infty f_\rho(iy)\, y^{s-\frac{k}{2}-1}\, dy \tag{3.7}$$

and, by definition, the function dies off exponentially fast as $y \to \infty$.

In general, a smooth function $f : \{x + iy \mid y > 0\} \to \mathbb{C}$ is called a *Maass form* of weight $k \in \mathbb{Z}$, level $N \in \mathbb{Z}$, and nebentype $\varepsilon : \mathbb{Z} \to \mathbb{C}$ if it satisfies the following properties:

1. *Eigenfunction of the non-Euclidean Laplacian.* $f(x + iy)$ satisfies the differential equation

$$\left\{ -y^2 \left(\frac{\partial^2}{\partial x^2} + \frac{\partial^2}{\partial y^2} \right) + i\, k\, y\, \frac{\partial}{\partial x} \right\} f(\tau) = \frac{k}{2} \left(1 - \frac{k}{2} \right) f(\tau) \tag{3.8}$$

2. *Exponential Decay.* For any complex s with $\mathrm{Re}(s) > 1$,

$$\lim_{y \to \infty} f(x + iy)\, y^{s-1} = 0 \tag{3.9}$$

3. *Transformation Property.* For all matrices in the group

$$\Gamma_0(N) = \left\{ \begin{pmatrix} a & b \\ c & d \end{pmatrix} \in \mathrm{Mat}_2(\mathbb{Z}) \middle| \, ad - bc = 1 \text{ and } c \text{ is divisible by } N \right\} \tag{3.10}$$

we have the identity

$$f\left(\frac{a\, \tau + b}{c\, \tau + d} \right) = \varepsilon(d) \left(\frac{c\, \overline{\tau} + d}{c\, \tau + d} \right)^{k/2} f(\tau) \tag{3.11}$$

4. If in addition we have

$$\int_0^1 f(\tau + x)\, dx = 0 \qquad \text{for all } \tau \in \{x + iy \mid y > 0\}; \tag{3.12}$$

we say that f is a *cusp* form. Otherwise, we call f an *Eisenstein series.*

For a given irreducible complex representation ρ which is ramified outside of a finite number of primes Σ, the series in (3.6) always satisfies the first two conditions. The condition in (3.12) is equivalent to $a_\rho(0) = 0$, which happens if and only if ρ is not the trivial representation. Hence, in order to invoke Theorem 3.3 it suffices to prove the transformation property. Langlands succeeded in proving this in many cases by proving a generalization of the Selberg Trace Formula. Unfortunately, there does not appear to be a way to use these ideas in the icosahedral case.

4. Constructing Examples of Icosahedral Representations

Not many examples satisfying Artin's Conjecture in the icosahedral case are known. Those that are may be generated by the following program, initiated by Joe Buhler [**3**] and furthered by Ian Kiming [**7**].

1. Construct an A_5-extension $K/\mathbb{Q}$ by considering quintics.
2. Consider the discriminant of K in order find possible conductors. This will uniquely specify the representation.
3. Construct weight 2 cusp forms by considering dihedral representations.
4. Divide by an Eisenstein series to find a weight 1 form.

While this program is straightforward, there are two computational barriers. First, finding the discriminant of an extension can be a tedious procedure. Unfortunately, the method above relies on this step in order to pinpoint the representation and to generate the weight 2 cusp forms. Second, division by Eisenstein series is much more difficult that it sounds. One must find the zeroes of the weight 2 cusp forms, the zeroes of the Eisenstein series, and then show that they occur at the same places.

Motivated by these difficulties, we ask

Can we generate Buhler's/Kiming's examples by using elliptic curves?

We modify the program above with this question in mind.

1. Construct an A_5-extension $K/\mathbb{Q}$ by considering quintics. Use classical results due to Klein to associate elliptic curves.
2. Construct the A_5-representations by considering the 5-torsion.
3. Construct weight 2 cusp forms associated to the elliptic curve.
4. Multiply by an Eisenstein series to find a weight 5 form.

4.1. Step #1: Relating A_5-Extensions to Elliptic Curves. Felix Klein showed how to associate an elliptic curve to a certain class of polynomials.

THEOREM 4.1 (Klein [**12**]). *Fix $q(x) = x^5 + A\,x^2 + B\,x + C$ as a polynomial with rational coefficients, and assume that $q(x)$ has Galois group A_5. Once one solves for j in the system*

$$\begin{aligned} A &= -\frac{20}{j}\left[2\,m^3 + 3\,m^2\,n + 432\,\frac{6\,m\,n^2 + n^3}{1728 - j}\right] \\ B &= -\frac{5}{j}\left[m^4 - 864\,\frac{3\,m^2\,n^2 + 2\,m\,n^3}{1728 - j} + 559872\,\frac{n^4}{(1728 - j)^2}\right] \\ C &= -\frac{1}{j}\left[m^5 - 1440\,\frac{m^3\,n^2}{1728 - j} + 62208\,\frac{15\,m\,n^4 + 4\,n^5}{(1728 - j)^2}\right] \end{aligned} \tag{4.1}$$

then every root of $q(x)$ can be expressed in terms of the 5-torsion on any elliptic curve E with $j = j(E) \in \mathbb{Q}(\sqrt{5})$.

As an example, consider the quintic $x^5 + 10\,x^3 - 10\,x^2 + 35\,x - 18$. This is not in the form of the principal quintic above, but after making the substitution

$$x \mapsto \frac{(1 + \sqrt{5})\,x + (10 - 30\sqrt{5})}{2\,x + (35 + 5\sqrt{5})} \tag{4.2}$$

the polynomial of interest is

$$x^5 - 125\left(185 + 39\sqrt{5}\right)x^2 - 6875\left(56 + 19\sqrt{5}\right)x - 625\left(10691 + 2225\sqrt{5}\right) \tag{4.3}$$

with ramified primes $\Sigma = \{2, \sqrt{5}\}$. One solves the equations above to find the elliptic curve

$$E_0: \quad y^2 = x^3 + (5 - \sqrt{5})\,x^2 + \sqrt{5}\,x; \qquad j(E_0) = 86048 - 38496\,\sqrt{5}. \tag{4.4}$$

4.2. Step #2: Constructing Icosahedral Representations. Once the ellipticcurve is found, one constructs the icosahedral representation by following a rather simple algorithm.

LEMMA 4.2 (Goins [**8**], Klute [**13**]). *Let E be an elliptic curve as constructed above, and Σ a set containing ramified primes. Then there exists an icosahedral representation ρ_E with L-series*

$$L_\Sigma(\rho_E, s) = \prod_{\mathfrak{p} \notin \Sigma} \left(1 - \frac{a_E(\mathfrak{p})}{\mathbb{N}\,\mathfrak{p}^s} + \frac{\omega_5(\mathbb{N}\,\mathfrak{p})}{\mathbb{N}\,\mathfrak{p}^{2s}}\right)^{-1} \tag{4.5}$$

where ω_5, a Dirichlet character modulo 5, is the nebentype; and $a_E(\mathfrak{p}) \in \mathbb{Q}(i, \sqrt{5})$ is the trace of Frobenius.

We explain how this works in the simplest case, when the elliptic curve is defined over $\mathbb{Q}(\sqrt{5})$. Given a prime number p, denote the prime ideal lying above p as

$$\mathfrak{p} = \left\{ a + \frac{-1+\sqrt{5}}{2}\,b \in \mathbb{Q}(\sqrt{5}) \,\middle|\, a,\, b \in \mathbb{Z};\ p \text{ divides } a^2 - a\,b - b^2 \right\} \tag{4.6}$$

The nebentype may be expressed as

$$\omega_5(\mathbb{N}\,\mathfrak{p}) = \left(\frac{5}{p}\right) \qquad \text{where} \qquad \mathbb{N}\,\mathfrak{p} = \begin{cases} p & \text{if } p \equiv 1, 4 \mod 5; \\ p^2 & \text{if } p \equiv 2, 3 \mod 5. \end{cases} \tag{4.7}$$

To calculate the trace of Frobenius, one would factor the polynomial

$$(x+3)^3\,(x^2 + 11\,x + 64) - j(E) \mod \mathfrak{p} \tag{4.8}$$

consult the table

Irred. Factors	Linear	Quadratics	Cubic	Quintic
$\dfrac{a_E(\mathfrak{p})^2}{\omega_5(\mathbb{N}\,\mathfrak{p})}$	4	0	1	$\left(\dfrac{-1 \pm \sqrt{5}}{2}\right)^2$

and finally decide upon which square root by the congruence

$$a_E(\mathfrak{p}) \equiv \mathbb{N}\,\mathfrak{p} + 1 - |\widetilde{E}(\mathbb{F}_\mathfrak{p})| \mod \left(2 - i,\, \sqrt{5}\right) \tag{4.9}$$

where $|\widetilde{E}(\mathbb{F}_\mathfrak{p})|$ is the number of points on the elliptic curve mod $\mathfrak{p}$.

As an application, we take a closer look at the elliptic curve associated to the polynomial $x^5 + 10\,x^3 - 10\,x^2 + 35\,x - 18$.

PROPOSITION 4.3. *Let E_0 be the elliptic curve $y^2 = x^3 + (5-\sqrt{5})\,x^2 + \sqrt{5}\,x$.*

1. *E_0 is isogeneous over $\mathbb{Q}(\sqrt{5}, \sqrt{-2})$ to each of its Galois conjugates. That is, E_0 is a $\mathbb{Q}$-curve.*
2. *There is a character χ_0 such that $\chi \otimes \rho_{E_0}$ is the base change of an icosahedral representation ρ with conductor $N(\rho) = 800$ and nebentype $\epsilon_\rho = \left(\frac{-1}{\cdot}\right)$. Specifically, ρ is the icosahedral Galois representation studied in* [**3**].

PROOF. We sketch the ideas. The L-series of the twisted representation $\chi \otimes \rho_{E_0}$ is defined as

$$L_\Sigma\,(\rho_{E_0},\, \chi,\, s) = \prod_{\mathfrak{p} \notin \Sigma} \left(1 - \frac{\chi(\mathfrak{p})\, a_{E_0}(\mathfrak{p})}{\mathbb{N}\,\mathfrak{p}^s} + \frac{\chi(\mathfrak{p})^2\, \omega_5(\mathbb{N}\,\mathfrak{p})}{\mathbb{N}\,\mathfrak{p}^{2s}}\right)^{-1} \tag{4.10}$$

If we were to find a character χ such that 1) it is unramified outside of Σ; 2) $\chi(\sigma\,\mathfrak{p})\, a_{E_0}(\sigma\,\mathfrak{p}) = \chi(\mathfrak{p})\, a_{E_0}(\mathfrak{p})$ for all $\sigma \in G_{\mathbb{Q}}$; and 3) $\chi(\mathfrak{p})^2\, \omega_5(\mathbb{N}\,\mathfrak{p}) = \epsilon_\rho(\mathbb{N}\,\mathfrak{p})$; then the L-series would be in the form

$$L_\Sigma\,(\rho_{E_0},\, \chi,\, s) = \prod_{\mathfrak{p} \notin \Sigma} \left(1 - \frac{\alpha(\mathbb{N}\,\mathfrak{p})}{\mathbb{N}\,\mathfrak{p}^s} + \frac{\epsilon_\rho(\mathbb{N}\,\mathfrak{p})}{\mathbb{N}\,\mathfrak{p}^{2s}}\right)^{-1} = L_\Sigma\left(\rho|_{\mathbb{Q}(\sqrt{(5)}},\, s\right) \tag{4.11}$$

for some representation ρ defined over $\mathbb{Q}$. Cleary ρ has nebentype ϵ_ρ and is unramified outside of Σ, so that ρ is the unique representation studied in [**3**]. It suffices to construct the character χ.

The elliptic curve E_0 is isogeneous to its congujates, which means

$$a_{E_0}(\sigma\,\mathfrak{p}) = \left(\frac{-2}{\mathbb{N}\,\mathfrak{p}}\right) a_{E_0}(\mathfrak{p}) \implies \begin{array}{c} \chi(\sigma\,\mathfrak{p}) = \left(\dfrac{-2}{\mathbb{N}\,\mathfrak{p}}\right) \chi(\mathfrak{p}) \\ \chi(\mathfrak{p})^2 = \omega_5(\mathbb{N}\,\mathfrak{p})^{-1} \left(\dfrac{-1}{\mathbb{N}\,\mathfrak{p}}\right) \end{array} \tag{4.12}$$

One constructs the character explicitly by considering the ideal of $\mathbb{Q}(\sqrt{5})$ lying above 40. □

We have shown the existence of the icosahedral representation in [**3**] without the worry of computing the discriminant of the splitting field. Moreover, using the algorithm above the coefficients can be calculated explicitly.

4.3. Step #3: Constructing Weight 2 Cusp Forms. We will exploit the fact that E_0 is isogenous to its Galois conjugates. While it is not necessary in general that E be a $\mathbb{Q}$-curve in order to find an icosahedral representation using the steps outlined in the previous subsection, we do need this fact in this specific case to work with cusp forms. Indeed, using the formulas in the previous subsection one can show that there is always a character such that the twisted icosahedral representation comes from $\mathbb{Q}$, but a priori there seems to be little evidence that the elliptic curve will always be isogenous to its conjugates.

PROPOSITION 4.4. *Let E_0 be the elliptic curve $y^2 = x^3 + (5-\sqrt{5})\,x^2 + \sqrt{5}\,x$ and $\Sigma = \{2,\, \sqrt{5}\}$. There is a cusp form f_0 of weight 2, level 160 such that*

$$L_\Sigma(\rho,\, s) \equiv L_\Sigma(f_0,\, s) \mod \left(2 - i,\, \sqrt{5}\right) \tag{4.13}$$

PROOF. We compute the discriminant of the elliptic curve to see that it has good reduction outside of Σ. It is straightforward to use the Modular Symbol Algorithm [**5**] to calculate coefficients and match a cusp form $f_0{}'$ over $\mathbb{Q}(\sqrt{5})$ which is also unramified outside Σ. By 4.3, the twist $\chi \otimes f_0{}'$ is Galois invariant so it must be the base change of a cusp form f_0 over $\mathbb{Q}$. We use the Modular Symbol Algorithm again to find that the level is 160. □

4.4. Step #4: Constructing Weight 5 Cusp Forms. Using an idea of Ken Ribet [**17**], we may strip 5 altogether from the level as long as we increase the weight.

THEOREM 4.5. *Let E_0 and ρ be as in 4.4. There is a cusp form f_1 of weight 5, level 32 and nebentype $\epsilon_\rho = \left(\frac{-1}{*}\right)$ such that*

$$L(\rho_0, s) \equiv L(f_0, s) \equiv L(f_1, s) \mod \left(2 - i, \sqrt{5}\right). \tag{4.14}$$

PROOF. As defined in [**18**], consider the ℓ-adic Eisenstein series

$$\mathcal{E}(\tau) = 1 + \sum_{n=1}^{\infty} a_n(\mathcal{E})\, q^n; \quad a_n(\mathcal{E}) = 2\, \frac{\sum_{d|n} \epsilon_\ell(d)^{-k}\, d^{k-1}}{L\left(\epsilon_\ell{}^{-k},\, 1-k\right)} \in \ell\, \mathbb{Z}_\ell. \tag{4.15}$$

where ϵ_ℓ is the cyclotomic character. By [**18**, Lemme 10], this has weight k, level ℓ, nebentype ϵ_ℓ^{-w}, and satisfies $\mathcal{E} \equiv 1\ (\ell)$. Setting $w = 3$ and $\ell = 5$, the product $f_0 \cdot \mathcal{E}$ has weight 5, level 160, and nebentypus ϵ_ρ so by [**6**, Lemme 6.11], there is a bona fide eigenform $f_1 \equiv f_0 \cdot \mathcal{E}$. Using the Modular Symbol Algorithm one more time we see that f_1 has level 32. □

References

[1] Emil Artin. *The collected papers of Emil Artin.* Addison–Wesley Publishing Co., Inc., Reading, Mass.-London, 1965. Edited by Serge Lang and John T. Tate.

[2] Richard Brauer. On Artin's L-series with general group characters. *Ann. of Math. (2)*, 48:502–514, 1947.

[3] Joe P. Buhler. *Icosahedral Galois representations.* Springer-Verlag, Berlin-New York, 1978. Lecture Notes in Mathematics, Vol. 654.

[4] Daniel Bump. *Automorphic forms and representations.* Cambridge University Press, Cambridge, 1997.

[5] J. E. Cremona. *Algorithms for modular elliptic curves.* Cambridge University Press, Cambridge, second edition, 1997.

[6] Pierre Deligne and Jean-Pierre Serre. Formes modulaires de poids 1. *Ann. Sci. École Norm. Sup. (4)*, 7:507–530 (1975), 1974.

[7] G. Frey, editor. *On Artin's conjecture for odd 2-dimensional representations.* Springer-Verlag, Berlin, 1994.

[8] Edray H. Goins. *Elliptic Curves and Icosahedral Galois Representations.* PhD thesis, Stanford University, 1999.

[9] Erich Hecke. *Lectures on Dirichlet series, modular functions and quadratic forms.* Vandenhoeck & Ruprecht, Göttingen, 1983. Edited by Bruno Schoeneberg, With the collaboration of Wilhelm Maak.

[10] Erich Hecke. *Analysis und Zahlentheorie.* Friedr. Vieweg & Sohn, Braunschweig, 1987. Vorlesung Hamburg 1920. [Hamburg lectures 1920], Edited and with a foreword by Peter Roquette.

[11] Kenneth Ireland and Michael Rosen. *A classical introduction to modern number theory.* Springer-Verlag, New York, second edition, 1990.

[12] Felix Klein. *Vorlesungen über das Ikosaeder und die Auflösung der Gleichungen vom fünften Grade.* Birkhäuser Verlag, Basel, 1993. Reprint of the 1884 original, Edited, with an introduction and commentary by Peter Slodowy.

[13] Annette Klute. Icosahedral Galois extensions and elliptic curves. *Manuscripta Math.*, 93(3):301–324, 1997.

[14] Serge Lang. *Algebraic number theory.* Springer-Verlag, New York, second edition, 1994.

[15] Robert P. Langlands. *Base change for* GL(2). Princeton University Press, Princeton, N.J., 1980.

[16] Robert P. Langlands. L-functions and automorphic representations. In *Proceedings of the International Congress of Mathematicians (Helsinki, 1978)*, pages 165–175. Acad. Sci. Fennica, Helsinki, 1980.

[17] Kenneth A. Ribet. Report on mod l representations of gal($\overline{\mathbf{q}}/\mathbf{q}$). In *Motives (Seattle, WA, 1991)*, pages 639–676. Amer. Math. Soc., Providence, RI, 1994.

[18] Jean-Pierre Serre. Formes modulaires et fonctions zêta p-adiques. pages 191–268. Lecture Notes in Math., Vol. 350, 1973.

[19] Jerrold Tunnell. Artin's conjecture for representations of octahedral type. *Bull. Amer. Math. Soc. (N.S.)*, 5(2):173–175, 1981.

Institute for Advanced Study, School of Mathematics, Olden Lane, Princeton, NJ 08540

E-mail address: goins@math.ias.edu

Contemporary Mathematics
Volume **275**, 2001

Chaoticity results for "join the shortest queue".

Carl Graham

ABSTRACT. In a large queuing network, each task is allocated at arrival a small number of queues and elects to join the shortest one. This introduces resource pooling based on limited information. The load is thus better balanced and servers are better utilized throughout the network; essentially, tasks find less tasks ahead of them. We consider the statistical mechanics limit, in which the network size goes to infinity. We prove propagation of chaos and chaoticity in equilibrium: the processes of the queue-lengths behave at the limit as independent processes, with law characterized by a nonlinear martingale problem. The limit stationary probability of the average queue size being above some threshold decays super-exponentially with the threshold.

1. Introduction

Packet-switched networks are used for the exchange of digital information in computer networks. Messages are cut up into packets with headers relating necessary information (source, destination, order in message ...). These packets are then transmitted through varied paths. They transit at intermediate nodes, at which they wait in buffers before being processed and sent on their way. The message must be reconstituted by reassembling the packets after transmission.

Quality of service and reliability are highly influenced by overall delay and queue overflow, which may lead to the loss of packets. This must be optimized in balance with the costs of construction, maintenance, and operation of the network, whose structure is often enormous, ill-known, and in constant evolution, as is the case for the Internet and the world-wide web. Network management is delocalized using sophisticated algorithms, and economy of scale effects are sought by pooling available resources so as to utilize them more efficiently.

Asymptotic studies are of interest in order to obtain qualitative and quantitative results. We refer to Graham [**2**] and the references therein for a wide discussion of loss and queuing networks in the large size limit. We shall concentrate in the sequel on results and techniques found in Graham [**3**].

We refer to the Appendices (Sections 5 and 6) for background material. General notations and terminology up to the links between Markov processes and martingale problems can be found in Section 5, more advanced material specific to chaoticity and nonlinear martingale problems in Section 6.

2000 *Mathematics Subject Classifications.* Primary 60K25, 60K35.

1.1. Join the shortest queue. An infinite-buffer queue with Poisson arrivals at rate ν and a single server at rate λ (the $M_\nu/M_\lambda/1/\infty$ queue) is ergodic if and only if $\nu/\lambda = \rho < 1$, and then has the geometric invariant law

$$(1.1) \qquad p\{k\} = (1-\rho)\rho^k \text{ or equivalently } p([k,\infty[) = \rho^k\,, \qquad k \in \mathbb{N}\,.$$

Hence in equilibrium the probability of having a large queue length, above some threshold, decreases exponentially with the threshold.

It is natural to try to balance the load over a network of such queues in order to better utilize its resources, using some simple delocalized scheme, and Vvedenskaya, Dobrushin and Karpelevich [**10**] introduce and study an interesting simple model. There are N identical infinite buffer queues, each with single server with rate λ. Tasks arrive at rate $N\nu$. Each task is allocated L queues uniformly at random among the N and joins the shortest one, ties being resolved uniformly. The arrival stream, allocations, and services are independent.

Turner [**9**] gives a theoretical and experimental study for this and related models. He notably provides a coupling showing that the load in the network is handled better and more evenly as L increases. Martin and Suhov [**6**] consider open Jackson networks in which such queuing networks replace the individual queues.

1.2. The statistical mechanics point of view. This is an L-body mean-field interacting system in which the interaction effect felt by one queue is expressed, in the large N limit, in terms of the empirical measure of L-tuples of queues.

Let $r_t^N(k)$ be the proportion of queues with at least $k \in \mathbb{N}$ waiting tasks at time $t \geq 0$. Vvedenskaya et al. [**10**] prove that if $\lim_{N\to\infty} r_0^N = u_0$ in law, for term wise convergence, then $\lim_{N\to\infty} r^N = u$ in law, for term-wise convergence uniformly on bounded time intervals, where u is the unique solution for a system of differential equations starting at u_0. This result is a functional law of large numbers (LLN) for initial conditions satisfying a LLN.

The network is ergodic if and only if $\rho = \nu/\lambda < 1$, with invariant law denoted by π^N. Vvedenskaya et al. [**10**] further prove that

$$(1.2) \qquad \lim_{N\to\infty} E_{\pi^N}(r_0^N(k)) = u^\rho(k) = \rho^{(L^k-1)/(L-1)}\,, \qquad k \in \mathbb{N}\,,$$

which decays super-exponentially in k. Comparing with (1.1), we see that the asymptotic probabilities of having large queue sizes are dramatically lower than for an i.i.d. system of queues.

We later explain methods yielding in particular the extensions of these results to path space summarized below. Let X_i^N be the length of queue $i \in \{1,\ldots,N\}$,

$$(1.3) \qquad \mu^N = \frac{1}{N}\sum_{i=1}^N \delta_{X_i^N}\,, \qquad \bar{X}_t^N = \frac{1}{N}\sum_{i=1}^N \delta_{X_i^N(t)}\,, \qquad \bar{X}^N = (\bar{X}_t^N)_{t\geq 0}\,.$$

The empirical measure μ^N with samples in $\mathcal{P}(\mathbb{D}(\mathbb{R}_+,\mathbb{N}))$ carries much more information than the marginal process $\bar{X}^N$ with sample paths in $\mathbb{D}(\mathbb{R}_+,\mathcal{P}(\mathbb{N}))$ or than r^N. We have $\bar{X}_t^N\{k\} = r_t^N(k) - r_t^N(k+1)$ and $r_t^N(k) = \bar{X}_t^N([k,\infty[)$.

We prove that if $\lim_{N\to\infty} \bar{X}_0^N = q$ in law for a deterministic $q \in \mathcal{P}(\mathbb{N})$ then $\lim_{N\to\infty} \mu^N = Q$ in probability, for the Skorohod topology, where Q uniquely solves a nonlinear martingale problem starting at q.

Also, we prove that in equilibrium $\lim_{N\to\infty} \mu^N = Q^\rho$ in probability, for the Skorohod topology, where Q^ρ uniquely solves a nonlinear martingale problem starting at q^ρ given by $q^\rho\{k\} = u^\rho(k) - u^\rho(k+1)$, $k \in \mathbb{N}$. This implies a functional LLN stronger than (1.2): in equilibrium, $\lim_{N\to\infty} \bar{X}^N = q^\rho$ in probability, for weak convergence uniformly on bounded time intervals (q^ρ denotes the constant process). Actually [**10**] proves but does not state a result very similar to this.

These LLN on path space yield limits for sample path quantities such as hitting times or extrema, which *cannot* be attained by functional laws of large numbers.

If $(X_i^N(0))_{1\le i\le N}$ is exchangeable, which is the case in equilibrium, we have chaoticity for the queuing network: for $k \ge 1$, $\lim_{N\to\infty} \mathcal{L}(X_1^N, \dots, X_k^N) = Q^{\otimes k}$ or $(Q^\rho)^{\otimes k}$, and the queues behave asymptotically as independent queues.

2. A coupling between networks with different number of choices

Turner [**9**], Theorem 4, constructs a coupling showing that the load is better balanced and handled more efficiently when the number of allocated queues is larger. For $L_0 \le L_1$ we couple two networks, called system 0 and system 1, allocated respectively L_0 and L_1 queues. For $\sigma \in \{0,1\}$ we denote quantities related to system σ by a superscript σ and consider the number of tasks which have at least m tasks queuing in front of them at time $t \ge 0$, defining the processes

$$c^{N,\sigma}(m) = N \sum_{k\ge m+1} r^{N,\sigma}(k) = \sum_{i=1}^{N} (X_i^{N,\sigma} - m)^+ , \qquad m \in \mathbb{N},\ \sigma \in \{0,1\}. \tag{2.1}$$

Note that $c_t^{N,\sigma}(0)$ is the total number of tasks in system σ at time $t \ge 0$.

We use a single Poisson process of rate $N\nu$ for arrivals for both systems. At each jump time, we choose uniformly $j_1^1 < \cdots < j_{L_1}^1$ among $1, \dots, N$ and then $j_1^0 < \cdots < j_{L_0}^0$ among $j_1^1, \dots, j_{L_1}^1$, and set $j^0 = j_{L_0}^0$ and $j^1 = j_{L_1}^1$. In system σ, we order the queues by *decreasing* length (ties are resolved with uniform probability), and let the task join the queue ranking j^σ in this order. Note that $j^0 \le j^1$.

We use a single Poisson process of rate $N\lambda$ for potential departures for both systems. At each jump time, we choose j uniformly in $\{1, \dots, N\}$. In system σ, we order again the queues by decreasing length, and remove a task from the j-th queue in this order if that queue is not empty.

THEOREM 2.1. *Let $N \ge 1$ and $L_1 \ge L_0$. If $c_0^{N,1}(m) \le c_0^{N,0}(m)$ for $m \in \mathbb{N}$ then*

$$c_t^{N,1}(m) \le c_t^{N,0}(m), \qquad m \in \mathbb{N},\ t \ge 0.$$

In particular, there are less tasks waiting in system 1 than in system 0.

PROOF. In the following, $\sigma \in \{0,1\}$, and τ is a jump time of the Poisson processes used for arrivals and departures. For any such time, we shall assume

$$c_{\tau-}^{N,1}(m) \le c_{\tau-}^{N,0}(m), \qquad m \in \mathbb{N}, \tag{2.2}$$

and prove that

$$c_{\tau}^{N,1}(m) \le c_{\tau}^{N,0}(m), \qquad m \in \mathbb{N}. \tag{2.3}$$

From (2.1) follows $c_t^{N,\sigma}(m) = N r_t^{N,\sigma}(m+1) + c_t^{N,\sigma}(m+1)$ for $m \ge 0$ and $t \ge 0$, thus if (2.2) holds then for any $n \ge 0$ we take $m = n-1$ and $m = n$ and obtain

$$c_{\tau-}^{N,1}(n) = c_{\tau-}^{N,0}(n) \Rightarrow r_{\tau-}^{N,1}(n) \le r_{\tau-}^{N,0}(n) \text{ and } r_{\tau-}^{N,1}(n+1) \ge r_{\tau-}^{N,0}(n+1). \tag{2.4}$$

At a departure time τ, let x^σ denote the respective lengths of the queues chosen for (potential) departures. A task will depart from system σ if and only if $x^\sigma > 0$, and then there will be one task less with exactly $x^\sigma - 1$ tasks in front of it, hence

$$c_\tau^{N,\sigma}(m) = c_{\tau-}^{N,\sigma}(m) - 1\,, \quad m < x^\sigma\,, \quad \text{and} \quad c_\tau^{N,\sigma}(m) = c_{\tau-}^{N,\sigma}(m)\,, \quad m \geq x^\sigma\,. \tag{2.5}$$

For the sake of contradiction, we assume that $c_\tau^{N,1}(n) > c_\tau^{N,0}(n)$ for some $n \geq 0$ (the negation of (2.3)). Then (2.2) and (2.5) imply that this is true if and only if

$$c_{\tau-}^{N,1}(n) = c_{\tau-}^{N,0}(n) \quad \text{and} \quad x^1 \leq n < x^0\,. \tag{2.6}$$

Now, if j in $\{1, \dots, N\}$ denotes the rank in decreasing order chosen for departures, then $Nr_{\tau-}^{N,\sigma}(x^\sigma + 1) < j \leq Nr_{\tau-}^{N,\sigma}(x^\sigma)$ and in particular $r_{\tau-}^{N,1}(x^1 + 1) < r_{\tau-}^{N,0}(x^0)$, which together with (2.6) and (2.4) yields

$$r_{\tau-}^{N,0}(n+1) \leq r_{\tau-}^{N,1}(n+1) \leq r_{\tau-}^{N,1}(x^1+1) < r_{\tau-}^{N,0}(x^0) \leq r_{\tau-}^{N,0}(n+1)$$

which is a contradiction. Thus (2.3) holds.

At an arrival time τ, let x^σ denote the lengths of the queues joined by either task. There is a new task in either system with x^σ tasks in front of it, hence

$$c_\tau^{N,\sigma}(m) = c_{\tau-}^{N,\sigma}(m) + 1\,, \quad m \leq x^\sigma\,, \quad \text{and} \quad c_\tau^{N,\sigma}(m) = c_{\tau-}^{N,\sigma}(m)\,, \quad m > x^\sigma\,. \tag{2.7}$$

Assuming $c_\tau^{N,1}(n) > c_\tau^{N,0}(n)$ for some $n \geq 0$, we obtain from (2.2) and (2.7) that this is true if and only if

$$c_{\tau-}^{N,1}(n) = c_{\tau-}^{N,0}(n) \quad \text{and} \quad x^0 < n \leq x^1\,. \tag{2.8}$$

If j^σ denotes the rank in decreasing order of the queue joined by the task in system σ, then $Nr_{\tau-}^{N,\sigma}(x^\sigma + 1) < j^\sigma \leq Nr_{\tau-}^{N,\sigma}(x^\sigma)$ and in particular $Nr_{\tau-}^{N,0}(x^0 + 1) < j^0 \leq j^1 \leq Nr_{\tau-}^{N,1}(x^1)$, which together with (2.8) and (2.4) yields

$$r_{\tau-}^{N,1}(n) \leq r_{\tau-}^{N,0}(n) \leq r_{\tau-}^{N,0}(x^0+1) < r_{\tau-}^{N,1}(x^1) \leq r_{\tau-}^{N,0}(n)$$

which is a contradiction. Thus (2.3) holds. The Theorem follows. □

We deduce from this coupling the ergodicity condition for the queuing network with selection among L queues.

THEOREM 2.2. *For $N \geq 1$, $(X_i^N)_{1\leq i\leq N}$ is ergodic if and only if $\rho < 1$, and then has a unique invariant law π^N. In equilibrium $(X_i^N)_{1\leq i\leq N}$ is exchangeable.*

PROOF. We take $L_0 = 1$ and $L_1 = L$ in the previous coupling. System 0 is an i.i.d. system of queues with arrival rates ν and service rates λ and is ergodic if and only if $\nu/\lambda = \rho < 1$. System 1 is the interacting system and is empty whenever system 0 is empty by Theorem 2.1, hence it is also ergodic for $\rho < 1$. Irreducibility implies uniqueness of the invariant law, and by a symmetry argument, the interacting system is exchangeable in equilibrium. The total number of customers in the system being larger than that of an appropriately defined single queue with arrival rate $N\nu$ and service rate $N\lambda$, the network is not ergodic for $\rho \geq 1$. □

3. Propagation of chaos for the queuing network

With the notations in Section 1.2, the process $(X_i^N)_{1\leq i\leq N}$ is Markov, and $\bar{X}_t^N\{k\}$ and $r_t^N(k)$ give the fraction of queues of length respectively *exactly equal* and *greater than or equal* to $k \in \mathbb{N}$ at time $t \geq 0$. The process $\bar{X}^N$ has sample paths in $\mathbb{D}(\mathbb{R}_+, \mathcal{P}(\mathbb{N}))$ and carries the same information as the process r^N with sample paths in $\mathbb{D}(\mathbb{R}_+, \mathcal{V})$, where

$$\mathcal{V} = \left\{(g(k))_{k\in\mathbb{N}} : g(0) = 1,\ g(k) \geq g(k+1),\ \lim_{k\to\infty} g(k) = 0)\right\}. \tag{3.1}$$

We consider the topology of term-wise convergence (product topology) on $\mathcal{V}$, which is then homeomorphic to $\mathcal{P}(\mathbb{N})$ with the weak topology.

Let $(n)_k = n(n-1)\cdots(n-k+1)$ for integer $n \geq k \geq 0$. We introduce the k-body empirical measures over all queues and over all queues other than $i \in \{1, \ldots, N\}$

$$(3.2)\quad \mu^{k,N} = \frac{1}{(N)_k} \sum_{\substack{i_1,\ldots,i_k=1\\ \text{distinct}}}^{N} \delta_{X^N_{i_1},\ldots,X^N_{i_k}}, \quad \mu^{k,N}_i = \frac{1}{(N-1)_k} \sum_{\substack{i_1,\ldots,i_k=1\\ \text{distinct},\ \neq i}}^{N} \delta_{X^N_{i_1},\ldots,X^N_{i_k}},$$

with samples in $\mathcal{P}(\mathbb{D}_{\mathbb{N}^k})$, and their marginal processes $\bar{X}^{k,N}$ and $\bar{X}^{k,N}_i$ with sample paths in $\mathbb{D}(\mathbb{R}_+, \mathcal{P}(\mathbb{N}^k))$. An important factorization lemma is at the core of L-body mean-field interaction. The combinatorial proof is straightforward.

LEMMA 3.1. *We have in total variation norm, uniformly on all outcomes,*

$$\left|\mu^{k,N} - (\mu^N)^{\otimes k}\right| = O(1/N) \text{ and } \left|\mu^{k,N}_i - (\mu^N)^{\otimes k}\right| = O(1/N).$$

For a function ϕ on $\mathbb{N}$, we set $\phi^+(x) = \phi(x+1) - \phi(x)$ and $\phi^-(x) = \phi(x-1) - \phi(x)$ (there will never be any ambiguities with the boundary point 0). Let

$$(3.3)\quad \chi : (x_1, \ldots, x_L) \in \mathbb{N}^L \longmapsto \frac{\mathbb{1}_{x_1 = \min\{x_1,\ldots,x_L\}}}{\sum_{i=1}^{L} \mathbb{1}_{x_i = \min\{x_1,\ldots,x_L\}}} \in \{0, 1/L, \ldots, 1/2, 1\}.$$

In the duality brackets between functions and measures below, the dot $\cdot$ represents the integration variable.

LEMMA 3.2. *Let $M^{\phi,i,N} = (M^{\phi,i,N}_t)_{t\geq 0}$ and $\varepsilon^{\phi,i,N} = (\varepsilon^{\phi,i,N}_t)_{t \geq 0}$ be given for a bounded function ϕ on $\mathbb{N}$ by*

$$\begin{aligned}
M^{\phi,i,N}_t &= \phi(X^N_i(t)) - \phi(X^N_i(0)) \\
&\quad - \int_0^t \left\{\nu L\langle \chi(X^N_i(s), \cdot), \bar{X}^{L-1,N}_{i,s}\rangle \phi^+(X^N_i(s)) \right.\\
&\quad \left. + \lambda \mathbb{1}_{X^N_i(s)\geq 1} \phi^-(X^N_i(s))\right\} ds \\
&= \phi(X^N_i(t)) - \phi(X^N_i(0)) \\
&\quad - \int_0^t \left\{\nu L\langle \chi(X^N_i(s), \cdot), (\bar{X}^N_s)^{\otimes L-1}\rangle \phi^+(X^N_i(s)) \right.\\
&\quad \left. + \lambda \mathbb{1}_{X^N_i(s)\geq 1} \phi^-(X^N_i(s))\right\} ds + \varepsilon^{\phi,i,N}_t.
\end{aligned} \tag{3.4}$$

Then $M^{\phi,i,N}$ and $M^{\phi,i,N}M^{\phi,j,N}$ are martingales for $i \neq j$, and we have $\varepsilon^{\phi,i,N}_t = t\|\phi\|_\infty O(1/N)$ uniformly on all outcomes.

PROOF. The martingale statements follow from Theorem 5.6 applied to the Markov process $(X^N_i)_{1\leq i\leq N}$ and the fact that there are no simultaneous jumps for X^N_i and X^N_j for $i \neq j$. The bound on $\varepsilon^{\phi,i,N}$ follows from Lemma 3.1. □

The nonlinear martingale problem. The law Q in $\mathcal{P}(\mathbb{D}_{\mathbb{N}})$ solves the nonlinear martingale problem starting at q iff $Q_0 = q$ and for any bounded function ϕ on $\mathbb{N}$,

(3.5)

$$M^\phi_t = \phi(X_t) - \phi(X_0) - \int_0^t \left\{\nu L\langle \chi(X_s, \cdot), Q_s^{\otimes L-1}\rangle \phi^+(X_s) + \lambda \mathbb{1}_{X_s \geq 1} \phi^-(X_s)\right\} ds$$

defines a Q-martingale.

We take the expectation, and use the simple result

$$L\left\langle \chi(\mathbf{x})\phi^+(x_1), Q_s^{\otimes L}(d\mathbf{x})\right\rangle = \left\langle \phi^+(\min\{\mathbf{x}\}), Q_s^{\otimes L}(d\mathbf{x})\right\rangle.$$

The nonlinear Kolmogorov equation and the differential system on $\mathcal{V}$. The process $(Q_t)_{t\ge 0}$ taking values in $\mathcal{P}(\mathbb{N})$ solves the nonlinear Kolmogorov equation starting at q iff $Q_0 = q$ and for any bounded function ϕ on $\mathbb{N}$, setting $\mathbf{x} = (x_1, \dots, x_L)$,

$$(3.6) \quad \langle \phi, Q_t\rangle = \langle \phi, Q_0\rangle + \int_0^t \left\{\nu\langle \phi^+(\min\{\mathbf{x}\}), Q_s^{\otimes L}(d\mathbf{x})\rangle + \lambda\langle \mathbb{I}_{\cdot\ge 1}\phi^-, Q_s\rangle\right\} ds\,.$$

Taking ϕ equal to $\mathbb{I}_{\{k\}}$, $k \in \mathbb{N}$, yields on $\mathcal{P}(\mathbb{N})$ an equivalent differential system for $Q_t\{k\}$, $k \in \mathbb{N}$, $t \ge 0$. We take instead $\phi_k = \mathbb{I}_{[k,\infty[}$, $k \ge 1$, in order to obtain the equivalent infinite system of scalar differential equations on $\mathcal{V}$ for $u_t(k) = Q_t([k,\infty[)$, $k \in \mathbb{N}$, $t \ge 0$, (note that $u(0) = 1$ identically)

$$(3.7) \qquad \dot u_t(k) = \nu\big(u_t(k-1)^L - u_t(k)^L\big) + \lambda\big(u_t(k+1) - u_t(k)\big)\,, \quad k \ge 1\,,$$

which corresponds to the systems (1.6) in Vvedenskaya et al. [**10**] and (1.4) in Martin and Suhov [**6**] (notations differ: their arrival rate is λ, their service rates are respectively 1 and μ). It is equivalent to solve (3.6) in $\mathcal{P}(\mathbb{N})$ or (3.7) in $\mathcal{V}$.

We are in the framework of Section 6 with $\mathcal{S} = \mathbb{N}$ and $\mathcal{A}$ given by (6.5) with

$$(3.8) \qquad J(q, x, dy) = \nu L\langle \chi(x,\cdot), q^{\otimes L-1}\rangle \delta_{x+1}(dy) + \lambda \mathbb{I}_{x\ge 1}\delta_{x-1}(dy)\,.$$

THEOREM 3.3. *There is a unique solution Q in $\mathcal{P}(\mathbb{D}_\mathbb{N})$ for the nonlinear martingale problem (3.5) starting at q. Its marginal process $(Q_t)_{t\ge 0}$ is the unique solution in $\mathcal{P}(\mathbb{N})$ for the nonlinear Kolmogorov equation (3.6), and $u = (u_t)_{t\ge 0}$ defined by $u_t(k) = Q_t([k,\infty[)$, $k \in \mathbb{N}$, is the unique solution in $\mathcal{V}$ for the system of differential equations (3.7) starting at u_0. The solutions depend continuously on their initial conditions.*

PROOF. We clearly have $|J(q,x)| \le \nu L + \lambda$. Since

$$|J(p,x) - J(q,x)| = \nu L|\langle \chi(x,\cdot), p^{\otimes L-1} - q^{\otimes L-1}\rangle| \le \nu L|p^{\otimes L-1} - q^{\otimes L-1}|$$

and $p^{\otimes L-1} - q^{\otimes L-1} = \sum_{i=1}^{L-1} p^{L-1-i} \otimes (p-q) \otimes q^{i-1}$, we have $|J(p,x) - J(q,x)| \le \nu L(L-1)|p-q|$. We conclude using Theorem 6.8. □

3.1. Propagation of chaos. We recall that $r_t^N(k) = \bar X_t^N([k,\infty[)$.

THEOREM 3.4. *Equivalently, assume that $(\bar X_0^N)_{N\ge 1}$ converges in law to q in $\mathcal{P}(\mathbb{N})$, or that $(r_0^N)_{N\ge 1}$ converges in law to u_0 in $\mathcal{V}$ (with $u_0(k) = q([k,\infty[)$, $k \in \mathbb{N}$). Then $(\mu^N)_{N\ge 1}$ converges in probability to the unique solution Q for the nonlinear martingale problem (3.5) starting at q, and if $(X_i^N(0))_{1\le i\le N}$ is exchangeable then $(X_i^N)_{1\le i\le N}$ is Q-chaotic. Moreover $(\bar X^N)_{N\ge 1}$ converges in probability to $(Q_t)_{t\ge 0}$ and $(r^N)_{N\ge 1}$ converges in probability to $u = (u_t)_{t\ge 0}$, for uniform convergence on bounded intervals, where $u_t(k) = Q_t([k,\infty[)$, $k \in \mathbb{N}$.*

PROOF. The law of μ^N is not changed when we permute the initial values, thus when we study it we may assume $(X_i^N(0))_{1\le i\le N}$ and hence $(X_i^N)_{1\le i\le N}$ exchangeable. We use a method developed by Sznitman [**8**] with three main steps:

1. Prove that $(\mathcal{L}(\mu^N))_{N\ge 1}$ is tight in $\mathcal{P}(\mathcal{P}(\mathbb{D}_\mathbb{N}))$.
2. Prove that the nonlinear martingale problem (3.5) is satisfied by any probability measure in the support of any accumulation point of $(\mathcal{L}(\mu^N))_{N\ge 1}$.
3. Prove that the nonlinear martingale problem (3.5) has at most one solution Q starting at any q in $\mathcal{P}(\mathbb{N})$.

Indeed, Step 2 and Step 3 imply that the only accumulation point of $(\mathcal{L}(\mu^N))_{N\geq 1}$ is δ_Q, and Step 1 then implies that $\lim_{N\to\infty}(\mathcal{L}(\mu^N))_{N\geq 1} = \delta_Q$ weakly. Theorem 6.1 yields the Q-chaoticity result. Theorem 6.2 concludes to the functional LLN.

We now execute the individual steps.

Step 1. It is equivalent to prove that $(\mathcal{L}(X_1^N))_{N\geq 1}$ is tight in $\mathcal{P}(\mathbb{D}_\mathbb{N})$, see Proposition 2.2 in Sznitman [**8**]. The jumps of X_1^N are included in those of a Poisson process A^N of rate $\nu L + \lambda$, and their size is bounded by 1, so the modulus of "continuity" in $\mathbb{D}_\mathbb{N}$ (Ethier and Kurtz [**1**] p. 122) of X_1^N is less than that of A^N. The law of A^N does not depend on $N \geq 1$, and since $(X_1^N(0))_{N\geq 1}$ converges and hence is tight, the basic tightness criterion in [**1**] p. 128 holds.

Step 2. Let Π^N the law of μ^N, $\Pi^\infty \in \mathcal{P}\big(\mathcal{P}(\mathbb{N})\big)$ an accumulation point. For any R in $\mathcal{P}(\mathbb{D}_\mathbb{N})$, $X \mapsto X_t$ is R-a.s. continuous for all t except perhaps in an at most countable subset D_R of $\mathbb{R}_+^*$ (Ethier and Kurtz [**1**] Lemma 7.7 p. 131) and further, it can be shown that $D = \{t \in \mathbb{R}_+ : \Pi^\infty(R : t \in D_R) > 0\}$ is at most denumerable (see the proof of Theorem 4.5 in Graham and Méléard [**5**]).

We take $0 \leq s_1 < \cdots < s_k \leq s < t$ outside D, $k \in \mathbb{N}$, and bounded ϕ and g on $\mathbb{N}$ and $\mathbb{N}^k$. We denote by E_R the expectation under $R \in \mathcal{P}(\mathbb{D}_\mathbb{N})$ on the canonical space, and define the Π^∞-a.s. continuous mapping

$$\begin{aligned} G : R \in \mathcal{P}(\mathbb{D}_\mathbb{N}) \quad \mapsto \quad & E_R\bigg(\bigg(\phi(X_t) - \phi(X_s) - \int_s^t \{\nu L\langle \chi(X_u,\cdot), R_u^{\otimes L-1}\rangle \phi^+(X_u) \\ & + \mathbb{1}_{X_u\geq 1}\phi^-(X_u)\}\, du\bigg) g(X_{s_1},\ldots,X_{s_k})\bigg). \end{aligned}$$

We set $g^{i,N} = g(X_i^N(s_1),\ldots,X_i^N(s_k))$, $Y_{s,t} = Y_t - Y_s$ for a process Y and use the notations in Lemma 3.2, and obtain using the exchangeability

$$\begin{aligned} \langle G^2, \Pi^N\rangle &= E(G(\mu^N)^2) = E\bigg(\Big(\frac{1}{N}\sum_{i=1}^N (M_{s,t}^{\phi,i,N} - \varepsilon_{s,t}^{\phi,i,N})g^{i,N}\Big)^2\bigg) \\ &= \frac{1}{N}E\Big(\big((M_{s,t}^{\phi,1,N} - \varepsilon_{s,t}^{\phi,1,N})g^{1,N}\big)^2\Big) \\ &\quad + \frac{N-1}{N}E\Big((M_{s,t}^{\phi,1,N} - \varepsilon_{s,t}^{\phi,1,N})g^{1,N}(M_{s,t}^{\phi,2,N} - \varepsilon_{s,t}^{\phi,2,N})g^{2,N}\Big) \end{aligned} \tag{3.9}$$

and then Lemma 3.2 implies that $\langle G^2, \Pi^N\rangle = O(1/N)$. Let lim denote the limit along the subsequence with limit Π^∞. The Fatou Lemma implies that $\langle G^2, \Pi^\infty\rangle \leq \lim\langle G^2, \Pi^N\rangle = 0$, and thus that $G(R) = 0$, Π^∞-a.s. Since this is holds for arbitrary bounded g and $0 \leq s_1 < \cdots < s_k \leq s \leq t$ not in the denumerable set D, we classically deduce that R solves the nonlinear martingale problem (3.5), and the continuity of $X \mapsto X_0$ implies that $R_0 = q$, Π^∞-a.s.

Step 3. This is given by Theorem 3.3. □

4. Chaoticity in equilibrium

The result of chaoticity in equilibrium will be obtained by a compactness-uniqueness method due to Whitt [**11**], essentially yielding the inversion of limits

$$\lim_{t\to\infty}\lim_{N\to\infty} = \lim_{N\to\infty}\lim_{t\to\infty}.$$

4.1. Long-time behavior of the limit. There is a classical bijection between

$$\mathcal{U} = \Big\{(g(k))_{k\in\mathbb{N}} : g(0) = 1,\ g(k) \geq g(k+1),\ \sum_{k\geq 0} g(k) < \infty)\Big\} \subset \mathcal{V} \tag{4.1}$$

and the set $\mathcal{P}_1(\mathbb{N})$ of probability measures with finite first moment. Theorem 1 in Vvedenskaya et al. [**10**] states a global attractiveness result in $\mathcal{U}$:

THEOREM 4.1. *Let $\rho = \nu/\lambda < 1$. Then there is a unique stationary solution in $\mathcal{U}$ for the differential system (3.7), equal at all times to*

$$u^\rho(k) = \rho^{(L^k-1)/(L-1)}, \quad k \in \mathbb{N}, \tag{4.2}$$

and the solution u of (3.7) starting at any u_0 in $\mathcal{U}$ is such that $\lim_{t\to\infty} u_t = u^\rho$. Equivalently, there is a unique stationary solution in $\mathcal{P}_1(\mathbb{N})$ for the nonlinear Kolmogorov equation (3.6), equal at all times to

$$q^\rho(k) = \rho^{(L^k-1)/(L-1)} - \rho^{(L^{k+1}-1)/(L-1)}, \quad k \in \mathbb{N}, \tag{4.3}$$

and the solution $(Q_t)_{t\geq 0}$ of (3.6) starting at q in $\mathcal{P}_1(\mathbb{N})$ is such that $\lim_{t\to\infty} Q_t = q^\rho$.

4.2. Limit theorems in equilibrium. We need some extension of Theorem 3.4 to initial conditions such that $(\bar{X}_0^N)_{N\geq 1}$ has a non-deterministic limit, but for the sake of simplicity we shall prove what needed in the following proof.

THEOREM 4.2. *Let $\rho = \nu/\lambda < 1$, and u^ρ and q^ρ be given in (4.2) and (4.3). In equilibrium, $(X_i^N)_{1\leq i\leq N}$ is Q^ρ-chaotic, where Q^ρ is the unique solution for the nonlinear martingale problem (3.5) starting at q^ρ, and X is in equilibrium under Q^ρ. Hence in equilibrium $(\bar{X}^N)_{N\geq 1}$ converges in probability to the constant process q^ρ and $(r^N)_{N\geq 1}$ converges in probability to the constant process u^ρ, for uniform convergence on bounded intervals.*

PROOF. We consider the coupling in Section 2 with $L_0 = 1$ and $L_1 = L$. Then system 0 is the i.i.d. system, system 1 the interacting system. We start both systems at the same initial value, with law the invariant law for the i.i.d. system. We have

$$E\big(X_1^{N,\sigma}(t)\big) = E\Big(\frac{1}{N}\sum_{i=1}^N X_i^{N,\sigma}(t)\Big) = \frac{1}{N}E(c_t^{N,\sigma}(0)), \quad N\geq 1, \ t\geq 0, \ \sigma\in\{0,1\},$$

and since $c_t^{N,1}(0) \leq c_t^{N,0}(0)$ for $t \geq 0$ by virtue of Theorem 2.1, we obtain

$$E\big(X_1^{N,1}(t)\big) \leq E\big(X_1^{N,0}(t)\big) = \rho(1-\rho)^{-1}$$

which is the mean for the law (1.1). Let $\mathcal{L}_{st}$, P_{st} and E_{st} denote the law, probability and expectation for the interacting network in equilibrium. We obtain

$$E_{st}\big(\langle x, \bar{X}_0^N(dx)\rangle\big) = E_{st}\big(X_1^N(0)\big) \leq \liminf_{t\to\infty} E\big(X_1^{N,1}(t)\big) \leq \rho(1-\rho)^{-1}, \quad N\geq 1,$$

using ergodicity and the Fatou Lemma. From these bounds follows that for all $N \geq 1$, $P_{st}(X_1^N(0) \geq K) \leq \rho(1-\rho)^{-1}K^{-1}$, hence $(\mathcal{L}_{st}(X_1^N(0)))_{N\geq 1}$ is tight and (equivalently) $(\mathcal{L}_{st}(\bar{X}_0^N))_{N\geq 1}$ is tight, and moreover that any point in the support of any accumulation point of $(\mathcal{L}_{st}(\bar{X}_0^N))_{N\geq 1}$ is in $\mathcal{P}_1(\mathbb{N})$, a.s.

Consider an accumulation point Π_0^∞ for $(\mathcal{L}_{st}(\bar{X}_0^N))_{N\geq 1}$ and a converging subsequence. Note that we do not know yet that Π_0^∞ is a Dirac mass, and we cannot use Theorem 3.4 before proving this fact. Nevertheless, we can prove exactly as for Theorem 3.4 that $(\mathcal{L}_{st}(\mu^N))_{N\geq 1} = (\Pi^N)_{N\geq 1}$ is tight along that subsequence, and that any law R in the support of any accumulation point Π^∞ in $\mathcal{P}(\mathcal{P}(\mathbb{D}_\mathbb{N}))$ satisfies the nonlinear martingale problem (3.5), in particular $(R_t)_{t\geq 0}$ solves the Kolmogorov equation (3.6). Moreover $\langle f, \Pi_t^N\rangle = E_{st}(f(\bar{X}_t^N)) = E_{st}(f(\bar{X}_0^N)) = \langle f, \Pi_0^N\rangle$ for any continuous bounded f, $N \geq 1$ and $t \geq 0$. We take the limit along a converging subsequence for $t \notin D$, and obtain that $\Pi_t^\infty = \Pi_0^\infty$, $t \geq 0$.

Take $\varepsilon > 0$ and an open neighborhood V of q^ρ. For $j \in \mathbb{N}$, let $\mathcal{P}_j$ be the set of all q in $\mathcal{P}_1(\mathbb{N})$ such that the solution for the Kolmogorov equation (3.6) starting at q is in V for all times $t \geq j$. The continuous dependence on q in Theorem 3.3 implies that $\mathcal{P}_j$ is measurable. Theorem 4.1 implies that $\mathcal{P}_1(\mathbb{N}) = \cup_j \mathcal{P}_j$, and since $\mathcal{P}_j \subset \mathcal{P}_{j+1}$, there is k such that $\Pi_0^\infty(\mathcal{P}_k) > 1 - \varepsilon$. Then $\Pi_0^\infty(V) = \Pi_k^\infty(V) = \Pi^\infty(R_k \in V) \geq \Pi^\infty(R_0 \in \mathcal{P}_k) = \Pi_0^\infty(\mathcal{P}_k) > 1 - \varepsilon$ and since V and ε are arbitrary, we have $\Pi_0^\infty(\{q^\rho\}) = 1$.

Hence Π_0^∞ is necessarily δ_{q^ρ} and by uniqueness $\lim_{N\to\infty} \Pi_0^N = \delta_{q^\rho}$ and we use Theorem 3.4 for the convergence results. Ethier and Kurtz [**1**] Lemma 9.1 p. 238 then implies that X is in equilibrium under Q^ρ. □

5. Appendix on probabilistic background

Ethier and Kurtz [**1**] gives most of this material. We use the duality brackets

$$\langle f, \mu \rangle = \langle f(x), \mu(dx) \rangle = \int f d\mu = \int f(x)\mu(dx) \,, \qquad f \in L^\infty(\mathcal{S}) \,, \ \mu \in \mathcal{M}_b(\mathcal{S}) \,.$$

5.1. Convergence in law and in probability. Let Ω be a **Polish** space (metric space, separable and complete), $\mathcal{F} = \mathcal{B}(\Omega)$ it Borel σ-algebra, P a probability measure. Let $\mathcal{S}$ be a Polish space, and X an $\mathcal{S}$-valued **random variable** (measurable function $\Omega \to \mathcal{S}$). For $B \subset \mathcal{S}$ we set $X^{-1}(B) = \{\omega \in \Omega : X(\omega) \in B\} = \{X \in B\}$. If B is in $\mathcal{B}(\mathcal{S})$ then $\{X \in B\}$ is in $\mathcal{B}(\Omega)$, and we define the **law** of X as the probability measure on $\mathcal{S}$, denoted by $\mathcal{L}(X)$ or $\mathcal{L}_X$, given by

$$\mathcal{L}_X(B) = P(X \in B) \,, \qquad B \in \mathcal{B}(\mathcal{S}) \,.$$

Then $(\mathcal{S}, \mathcal{B}(\mathcal{S}), \mathcal{L}_X)$ is a probability space, called the image of $(\Omega, \mathcal{F}, P)$ by X.

Let C_b denote the set of continuous bounded functions. If P^n, $n \geq 1$, and P^∞ are probability measures on $\mathcal{S}$, we say that $(P^n)_{n\geq 1}$ converges **weakly** to P iff

$$\lim_{n\to\infty} \langle f, P^n \rangle = \langle f, P^\infty \rangle \,, \qquad \forall f \in C_b(\mathcal{S}, \mathbb{R}) \,.$$

This convergence can be metrized in such a way that the set $\mathcal{P}(\mathcal{S})$ of probability measures on $\mathcal{S}$ is a Polish space. The Prokhorov Theorem gives a criterion for **relative compactness** (containment in a compact set, hence existence of converging subsequences) for a sequence of probability measures in this space : $(P^n)_{n\geq 1}$ is relatively compact if and only if (iff) it is **tight**, which means by definition

$$\forall \varepsilon > 0 \,, \ \exists \text{ a compact set } K_\varepsilon \text{ such that } P^n(\mathcal{S} - K_\varepsilon) < \varepsilon \,, \ \forall n \geq 1 \,. \tag{5.1}$$

If X^n, $n \geq 1$, and X^∞ are random variables with values in $\mathcal{S}$, with laws P^n, $n \geq 1$, and P, we say that $(X^n)_{n\geq 1}$ converges **in law** to X if and only if $(P^n)_{n\geq 1}$ converges weakly to P, or equivalently

$$\lim_{n\to\infty} E(f(X^n)) = E(f(X^\infty)) \,, \qquad \forall f \in C_b(\mathcal{S}, \mathbb{R}) \,.$$

We say that $(X^n)_{n\geq 1}$ is **tight** if $(P^n)_{n\geq 1}$ is tight; in (5.1) we may replace $P^n(\mathcal{S}-K_\varepsilon)$ by $P(X^n \notin K_\varepsilon)$. Note that these are notions on the probability laws, and not directly of the random variables considered as functions on a probability space.

A notion of convergence of random variables, considered as functions on the space Ω with metric d, is as follows. Let X^n, $n \geq 1$, and X^∞ be defined on Ω. Then $(X^n)_{n\geq 1}$ converges **in probability** to X if and only if

$$\forall \varepsilon > 0 \,, \quad \lim_{n\to\infty} P(d(X^n, X^\infty) > \varepsilon) = 0 \,.$$

It is simple to see that convergence in probability implies convergence in law. The converse is not true in general, but is true if X^∞ is deterministic (a.s. constant).

5.2. The Skorohod space of sample paths, and the canonical probability space. Stochastic processes model random evolutions in continuous time taking place in some Polish space $\mathcal{S}$, and can be considered as random variables with values (called sample paths) in an appropriate subset of $\mathcal{S}^{\mathbb{R}_+}$. A quite general choice is the Skorohod space $\mathbb{D}_\mathcal{S} = \mathbb{D}(\mathbb{R}_+, \mathcal{S})$ of right-continuous with left-hand limits paths $x : \mathbb{R}_+ \to \mathcal{S}$. The Skorohod metric is a complicated modification of the metric of uniform convergence on bounded time-intervals, such that $\mathbb{D}_\mathcal{S}$ is Polish.

Since the topology is separable, the Borel σ-algebra is the product σ-algebra, hence if $X = (X_t)_{t\geq 0}$ is a process with sample paths in $\mathbb{D}_\mathcal{S}$ then each X_t is a random variable with values in $\mathcal{S}$. If $P = \mathcal{L}(X) \in \mathcal{P}(\mathbb{D}_\mathcal{S})$ then $P_t = \mathcal{L}(X_t) \in \mathcal{P}(\mathcal{S})$ are called the marginals of P (at time $t \geq 0$). Essentially

$$P \in \mathcal{P}(\mathbb{D}(\mathbb{R}_+, \mathcal{S})) \mapsto (P_t)_{t\geq 0} \in \mathbb{D}(\mathbb{R}_+, \mathcal{P}(\mathcal{S}))$$

is a projection in which a lot of information is lost, and distinct laws on $\mathbb{D}_\mathcal{S}$ may well share the same marginals $(P_t)_{t\geq 0}$. Events determined by the state of the process at more than one time (hitting times, sojourn times, maximal values ...) cannot be given a probability using only $(P_t)_{t\geq 0}$. The law $P \in \mathcal{P}(\mathbb{D}_\mathcal{S})$ of X is characterized by the collection of the finite-dimensional marginals

$$P_{t_1\ldots t_k} = \mathcal{L}(X_{t_1}, \ldots, X_{t_k}) \in \mathcal{P}(\mathcal{S}^k), \qquad k \geq 1, \; 0 \leq t_1 < \cdots < t_k < \infty.$$

Any x in $\mathbb{D}(\mathbb{R}_+, \mathcal{S})$ can be obtained from its values on any dense **countable** set of $\mathbb{R}_+$ by taking appropriate limits from the right. This explains why the product σ-algebra is a reasonably large σ-algebra on $\mathbb{D}(\mathbb{R}_+, \mathcal{S})$ (it is not on $\mathcal{S}^{\mathbb{R}_+}$) and why we can find a reasonably strong topology which is separable.

Another important technical point is that the Skorohod topology is such that $x \in \mathbb{D}_\mathcal{S} \mapsto x(t)$ is **not** continuous. Nevertheless, if $(x_n)_{n\geq 0}$ converges to x in $\mathbb{D}_\mathcal{S}$ then $(x_n(t))_{n\geq 0}$ converges to $x(t)$ in $\mathcal{S}$ for any t in the **continuity set** of x, which is the complement in $\mathbb{R}_+$ of an at most **countable** set.

This issue of **separability** is essential in many ways. For instance there are good compactness criteria, of the Ascoli type, on $\mathbb{D}_\mathcal{S}$, which yield in turn good relative compactness criteria for weak convergence using the Prokhorov theorem. This can be used to prove convergence, notably weak convergence on $\mathcal{P}(\mathbb{D}_\mathcal{S})$, by **compactness-uniqueness** methods.

We may try to construct a process X with sample paths in $\mathbb{D}_\mathcal{S}$, defined on some probability space $(\Omega, \mathcal{F}, P)$, and satisfying some specified random evolution, in order to study its law. Since $(\mathbb{D}_\mathcal{S}, \mathcal{B}(\mathbb{D}_\mathcal{S}), \mathcal{L}_X)$ is a probability space, an efficient way of doing so is to consider the **canonical probability space** $(\mathbb{D}_\mathcal{S}, \mathcal{B}(\mathbb{D}_\mathcal{S}))$ with the **canonical process** $X = (X_t)_{t\geq 0}$ given by $X_t(\omega) = \omega_t$ for $\omega = (\omega_t)_{t\geq 0}$ in $\mathbb{D}_\mathcal{S}$. We then try to construct and study **directly** P on the canonical space so that X satisfies the random evolution, and automatically $P = \mathcal{L}(X)$.

5.3. Conditional expectations. Restricted information can be coded by the σ-algebra generated by the events in $(\Omega, \mathcal{F}, P)$ which are accessible to us. For instance if Y is an $\mathcal{S}$-valued random variable and if we can observe its outcomes, then we gain knowledge on the σ-algebra generated by Y, defined by

$$\sigma(Y) = \{\{Y \in B\} : B \in \mathcal{B}(\mathcal{S})\}.$$

We state a simple result stressing this fact.

LEMMA 5.1. *(Doob Lemma.) On $(\Omega, \mathcal{F}, P)$, let Y be a random variable with values in $\mathcal{S}$, and X be a $\sigma(Y)$-measurable random variable with values in $\mathcal{S}'$. Then there is a $\mathcal{L}_Y$-a.s. unique measurable function $h : \mathcal{S} \to \mathcal{S}'$ such that*

$$X(\omega) = h(Y(\omega)), \qquad \omega \in \Omega.$$

This is often written $X = h(Y)$, but we explicited the ω to emphasize the fact that h depends on the (functions) X and Y, but not on ω.

We consider a $\mathbb{R}^d$-valued random variable X, and some sub-σ-algebra $\mathcal{R}$ of $\mathcal{F}$ representing the restricted information which is available to us (usually $\mathcal{R} = \sigma(Y)$ for some observable random variable Y). We want to express the "best" prediction (or approximation) of X "given" this knowledge.

THEOREM 5.2. *Let a probability space $(\Omega, \mathcal{F}, P)$ and a sub-σ-algebra $\mathcal{R}$ of $\mathcal{F}$ be given. For any $\mathbb{R}^d$-valued X in $L^1(\Omega, \mathcal{F}, P)$ there exists an a.s. unique random variable $E(X \mid \mathcal{R})$, called the* **conditional expectation of X given $\mathcal{R}$**, *such that:*

1. *$E(X \mid \mathcal{R})$ belongs to $L^1(\Omega, \mathcal{R}, P)$ (is $\mathcal{R}$-measurable and integrable).*
2. *For any Z in $L^\infty(\Omega, \mathcal{R}, P)$, we have $E(ZX) = E(ZE(X \mid \mathcal{R}))$.*

If X is in $L^2(\Omega, \mathcal{F}, P)$ then $E(X \mid \mathcal{R})$ is the **orthogonal projection** *of X onto $L^2(\Omega, \mathcal{R}, P)$, and Property 2 is valid for all Z in $L^2(\Omega, \mathcal{R}, P)$ and corresponds to the characteristic property of the orthogonal projection.*

When $\mathcal{R} = \sigma(Y)$ for some random variable Y, we denote $E(X \mid \mathcal{R})$ by $E(X \mid Y)$ and call it the **conditional expectation of X given Y**. By the Doob Lemma (Lemma 5.1), there is some $\mathcal{L}_Y$-a.s. unique function h such that

$$E(X \mid Y)(\omega) = h(Y(\omega)), \qquad \omega \in \Omega,$$

and the **conditional expectation of X given that Y is equal to a** is given by

$$E(X \mid Y = a) = h(a), \qquad a \in Y(\Omega).$$

Note that $a \mapsto E(X \mid Y = a)$ is defined $\mathcal{L}_Y$-a.s., and that for $P(Y = a) \neq 0$ we have

$$E(X \mid Y = a) = \frac{E(X \mathbb{1}_{Y=a})}{P(Y = a)}.$$

For an event A we set $P(A \mid \mathcal{R}) = E(\mathbb{1}_A \mid \mathcal{R})$ and $P(A \mid Y = a) = E(\mathbb{1}_A \mid Y = a)$, respectively the **conditional probability** of A given $\mathcal{R}$ and given $Y = a$.

The main properties of the conditional expectation are essentially those of the expectation and those of an orthogonal projection. In particular:

THEOREM 5.3. *Under the assumptions and notations of Theorem 5.2:*

1. *If $X \geq 0$ a.s, then $E(X \mid \mathcal{R}) \geq 0$. (Positivity)*
2. *For $a, b \in \mathbb{R}$ and $X, Y \in L^1(\Omega, \mathcal{F}, P)$, we have $E(aX + bY \mid \mathcal{R}) = aE(X \mid \mathcal{R}) + bE(Y \mid \mathcal{R})$. (Linearity)*
3. *$\|E(X \mid \mathcal{R})\|_p \leq \|X\|_p$ for $1 \leq p \leq \infty$. (Contraction in the L^p spaces)*
4. *If X is $\mathcal{R}$-measurable then $E(X \mid \mathcal{R}) = X$.*
5. *For any $\mathcal{R}$-measurable Z such that ZX is in L^1, we have $E(ZX \mid \mathcal{R}) = ZE(X \mid \mathcal{R})$.*
6. *If $\mathcal{S}$ is a sub-algebra of $\mathcal{R}$, then $E(E(X \mid \mathcal{R}) \mid \mathcal{S}) = E(X \mid \mathcal{S})$.*
7. *We have $E(X \mid \{\emptyset, \Omega\}) = E(X)$. In particular, $E(E(X \mid \mathcal{R})) = E(X)$.*
8. *If X and Y are independent and integrable, then $E(X \mid Y) = E(X)$.*

5.4. Markov processes, their generators, and martingale problems. Markov processes model phenomena for which *the probabilistic prediction of the future evolution after a time $s \geq 0$, given knowledge of all the past before s, actually depends only on the present at time s.*

This prediction may depend on both the position at time s and on the time s itself, in which case we speak of **inhomogeneous** Markov processes. We focus on the case when the prediction depends only on the position, in which we speak of **homogeneous** Markov processes, and shall see that most notions and results can be generalized with proper thought. We often drop the term "homogeneous".

Let $(\Omega, \mathcal{F}, P)$ be given. A **filtration** is an increasing family $(\mathcal{F}_t)_{t\geq 0}$ of sub-σ-fields where $\mathcal{F}_t$ models the information available to an observer up to time t,

$$\mathcal{F}_s \subset \mathcal{F}_t \subset \mathcal{F}\,, \qquad 0 \leq s \leq t < \infty\,.$$

If X is a process on $(\Omega, \mathcal{F})$, then its **proper filtration** $(\mathcal{F}_t^X)_{t\geq 0}$ is given by

$$\mathcal{F}_t^X = \sigma((X_s)_{0\leq s\leq t})$$

corresponding to the information obtained by observing the process up to time t.

We consider a **filtered probability space** $(\Omega, \mathcal{F}, (\mathcal{F}_t)_{t\geq 0}, P)$. A process X is **adapted to the filtration** $(\mathcal{F}_t)_{t\geq 0}$ if and only if for all $t \geq 0$, X_t is $\mathcal{F}_t$-measurable (equivalently $\mathcal{F}_t^X \subset \mathcal{F}_t$). Processes will be **adapted** if not stated otherwise.

The **canonical filtered probability space** is the canonical space $\mathbb{D}_{\mathcal{S}}$ with the canonical filtration $\mathcal{F}^X$ generated by the canonical process X. Since we are interested in the **laws** of Markov processes, we define and study them directly on the canonical path space.

DEFINITION 5.4. The law $P \in \mathcal{P}(\mathbb{D}_{\mathcal{S}})$ is an **inhomogeneous Markov law** iff

$$E(\phi(X_{s+t}) \,|\, \mathcal{F}_s) = E(\phi(X_{s+t}) \,|\, X_s)\,, \qquad \forall t, s \geq 0\,,\ \phi \in L^\infty(\mathcal{S})\,. \tag{5.2}$$

The family $P_x \in \mathcal{P}(\mathbb{D}_{\mathcal{S}})$, for x in $\mathcal{S}$, such that $P_x(X_0 = x) = 1$, is a family of **(homogeneous) Markov laws** if and only if P_x satisfies (5.2) for $x \in \mathcal{S}$ and

$$E_x(\phi(X_{s+t}) \,|\, X_s = y) = E_y(\phi(X_t))\,, \qquad \forall t, s \geq 0\,,\ \phi \in L^\infty(\mathcal{S})\,,\ x, y \in \mathcal{S}\,. \tag{5.3}$$

The family $(T_t)_{t\geq 0}$ of linear positive contraction operators

$$T_t : \phi \in L^\infty(\mathcal{S}) \mapsto T_t\phi \in L^\infty(\mathcal{S})\,, \quad T_t\phi(x) = E_x(\phi(X_t)) = \langle \phi, P_{x,t}\rangle\,, \tag{5.4}$$

satisfying $T_t 1 = 1$ and the **semi-group property**

$$T_0 = I_d\,, \quad T_{s+t} = T_s \circ T_t\,, \qquad t, s \geq 0\,, \tag{5.5}$$

is called the **semi-group** of the Markov process. Its **generator** $\mathcal{A}$ with **domain** $\mathcal{D}(\mathcal{A})$ is defined by $\mathcal{A}\phi = \lim_{t\to 0}(T_t\phi - \phi)/t$ with $\mathcal{D}(\mathcal{A})$ the set of $\phi \in L^\infty(\mathcal{S})$ such that the limit exists, and using probabilistic notations

$$\mathcal{A}\phi(x) = \lim_{t\to 0} \frac{E_x(\phi(X_t)) - \phi(x)}{t}\,, \qquad \phi \in \mathcal{D}(\mathcal{A})\,,\ x \in \mathcal{S}\,. \tag{5.6}$$

The Hille-Yosida theory explains how to interpret conversely T_t as $e^{t\mathcal{A}}$, which can be done simply by a series expansion if $\mathcal{A}$ is a bounded operator on $L^\infty(\mathcal{S})$.

For discrete $\mathcal{S}$, a Markov process with sample paths in $\mathbb{D}_{\mathcal{S}}$ stays at any x in $\mathcal{S}$ for an exponential random variable T of parameter $\lambda(x) \geq 0$ (i.e., $P(T > t) = e^{-\lambda(x)t}$)

and then jumps to a point $y \neq x$ according to a probability kernel $\pi(x, dy)$. Setting $J(x, dy) = \lambda(x)\pi(x, dy)$ the generator is given by

$$\mathcal{A}\phi(x) = \lambda(x)\left(\int_{\mathcal{S}} \phi(y)\,\pi(x, dy) - \phi(x)\right) = \int_{\mathcal{S}} \{\phi(y) - \phi(x)\} J(x, dy)\,. \tag{5.7}$$

We have from the definition of the generator that

$$E_x(\phi(X_t)) = \phi(x) + \int_0^t E_x(\mathcal{A}\phi(X_s))\, ds\,, \qquad t \geq 0\,,\ \phi \in \mathcal{D}(\mathcal{A})\,,\ x \in \mathcal{S}\,. \tag{5.8}$$

We often denote P_x or $P_{x,t}$ simply by P or P_t with initial condition $P_0 = \delta_x$, and interpret (5.8) as the linear evolution equation for the marginal process $(P_t)_{t\geq 0}$

$$\langle \phi, P_t \rangle = \langle \phi, P_0 \rangle + \int_0^t \langle \mathcal{A}\phi, P_s \rangle\, ds\,, \qquad t \geq 0\,,\ \phi \in \mathcal{D}(\mathcal{A})\,, \tag{5.9}$$

corresponding to a weak formulation for the equation $\partial_t P_t = \mathcal{A}^* P_t$ in which $\mathcal{A}^*$ is the formal **dual** or **adjoint** operator of $\mathcal{A}$. This equation is called the (forward) **Kolmogorov equation** for the Markov process. If $\mathcal{S}$ is discrete, it corresponds to a system (indexed by $\mathcal{S}$) of scalar differential equations evolving in the simplex.

The generator expresses the trend of the Markov process, with a remainder which is a martingale, as we shall now see.

DEFINITION 5.5. A process M on $(\Omega, \mathcal{F}, (\mathcal{F}_t)_{t\geq 0}, P)$ is a **martingale** iff

1. It is adapted and integrable: $M_t \in L^1(\Omega, \mathcal{F}_t, P)$, for all $t \geq 0$.
2. It is conditionally constant: $E(M_t \,|\, \mathcal{F}_s) = M_s$, for all $0 \leq s \leq t$.

There are many consequences to this definition. Let us just note that

$$E(M_t \,|\, \mathcal{F}_s) = M_s \;\Leftrightarrow\; E(M_t - M_s \,|\, \mathcal{F}_s) = 0 \;\Rightarrow\; E(M_t) = E(M_s)\,, \quad 0 \leq s \leq t\,.$$

THEOREM 5.6. *(Dynkin formula.) Let P_x, x in $\mathcal{S}$, be a family of Markov laws with generator $\mathcal{A}$ with domain $\mathcal{D}(\mathcal{A})$. Set*

$$\phi(X_t) - \phi(X_0) - \int_0^t \mathcal{A}\phi(X_s)\, ds = M_t^\phi\,, \qquad t \geq 0\,,\ \phi \in \mathcal{D}(\mathcal{A})\,.$$

For any x in $\mathcal{S}$, $(M_t^\phi)_{t\geq 0}$ is a martingale under P_x for the canonical filtration.

Since $M_0^\phi = 0$ we have $E_x(M_t^\phi) = 0$, $t \geq 0$, which yields the Kolmogorov equation (5.8), (5.9), in which the random fluctuations are averaged out. There is much more information about sample path behavior in the definition of a martingale than this mean-zero property. We may now characterize a Markov process in a way inspired by the Kolmogorov equation.

DEFINITION 5.7. Let $\mathcal{A}$ be a linear operator defined on $\mathcal{C} \subset L^\infty(\mathcal{S})$, and set

$$\phi(X_t) - \phi(X_0) - \int_0^t \mathcal{A}\phi(X_s)\, ds = M_t^\phi\,, \qquad t \geq 0\,,\ \phi \in \mathcal{C}\,.$$

We say that $P_x \in \mathbb{D}_{\mathcal{S}}$ solves the **martingale problem** for $\mathcal{A}$ starting at x if $P_x(X_0 = x) = 1$ and if $(M_t^\phi)_{t\geq 0}$ is a martingale under P_x for the canonical filtration.

We do not seek for the maximum generality, and assume that we are in a situation in which we know beforehand that the sample paths of solutions to the martingale problem on product space $\mathcal{S}^{\mathbb{R}_+}$ are a.s. in the Skorohod space $\mathbb{D}_{\mathcal{S}}$. The following results are extracted from Ethier and Kurtz [**1**], pp. 182–188.

THEOREM 5.8. *Assume that for each $x \in \mathcal{S}$ there exists a unique solution P_x for the martingale problem for $\mathcal{A}$ starting at x. Then (P_x), x in $\mathcal{S}$, is a family of Markov laws with generator extending $\mathcal{A}$. Conversely, if (P_x), x in $\mathcal{S}$, is a family of Markov laws with generator $\mathcal{A}$, and $\mathcal{C} \subset \mathcal{D}(\mathcal{A})$ is a separating set, then it is the unique solution for the martingale problem in Definition 5.7.*

We take for $\mathcal{C}$ a convenient set of sufficiently regular functions, sufficiently large for uniqueness to hold. Choices usually contain the set of smooth functions with compact support C_K^∞. We may take $\mathcal{C} = L^\infty(\mathcal{S})$ if $\mathcal{A}$ is a bounded operator.

REMARK 5.9. If X is an inhomogeneous Markov process on $\mathcal{S}$, then $(t, X_t)_{t\geq 0}$ is a homogeneous Markov process on $\mathbb{R}_+ \times \mathcal{S}$. Its generator acts on functions $\phi(t,x)$ which are differentiable in the variable $t \in \mathbb{R}_+$ in the form $(\partial_t + \mathcal{A}_t)\phi(t,x)$, where $\mathcal{A}_t$ is a generator acting only on the $x \in \mathcal{S}$ variable ($\mathcal{A}_t$ depends on t if and only if the process is homogeneous). This allows to generalize most results for homogeneous processes to inhomogeneous ones, see Ethier and Kurtz [**1**] pp. 221-223.

6. Appendix on chaoticity and nonlinear martingale problems

Let $\mathcal{S}$ be a Polish space (in the paper, one of $\mathbb{N}$, $\mathbb{N}^k$, $\mathbb{D}(\mathbb{R}_+, \mathbb{N})$, $\mathbb{D}(\mathbb{R}_+, \mathbb{N}^k)$, or the spaces of probability measures on these spaces).

6.1. Chaoticity and laws of large numbers.

For Q in $\mathcal{P}(\mathcal{S})$, the sequence of random variables $(X_i^N)_{1\leq i\leq N}$ on $\mathcal{S}^N$ is **Q-chaotic** if and only if (iff)

$$\forall k \geq 1\,, \quad \lim_{N\to\infty} \mathcal{L}(X_1^N, \dots, X_k^N) = Q^{\otimes k} \quad \text{weakly in } \mathcal{P}(\mathcal{S}^k), \tag{6.1}$$

in short, iff any fixed finite collection of variables is asymptotically i.i.d. of law Q. If $(X_i^N)_{1\leq i\leq N}$ are processes defined in such a way that $(X_i^N)_{1\leq i\leq N}$ is chaotic whenever $(X_i^N(0))_{1\leq i\leq N}$ is, then we say that there is **propagation of chaos**.

The empirical measure on path space, with samples in $\mathcal{P}(\mathbb{D}_\mathcal{S})$, is defined as

$$\mu^N = \frac{1}{N}\sum_{i=1}^N \delta_{X_i^N} \tag{6.2}$$

and is an important macroscopic object. For clarity we denote by

$$\bar{X}^N = (\bar{X}_t^N)_{t\geq 0}\,, \quad \bar{X}_t^N = \mu_t^N = \frac{1}{N}\sum_{i=1}^N \delta_{X_i^N(t)}\,, \tag{6.3}$$

its process of time marginals, with sample paths in $\mathbb{D}(\mathbb{R}_+, \mathcal{P}(\mathcal{S}))$. The following classical result is given and proved in Proposition 2.2 in Sznitman [**8**] p. 177. We recall that $(X_i^N)_{1\leq i\leq N}$ is **exchangeable** iff its law is invariant under permutations.

THEOREM 6.1. *Let Q be in $\mathcal{P}(\mathcal{S})$. If $(X_i^N)_{1\leq i\leq N}$ is Q-chaotic, then we have $\lim_{N\to\infty} \mathcal{L}(\mu^N) = \delta_Q$ weakly in $\mathcal{P}(\mathcal{P}(\mathcal{S}))$. If $(X_i^N)_{1\leq i\leq N}$ is exchangeable for $N \geq 1$ and $\lim_{N\to\infty} \mathcal{L}(\mu^N) = \delta_Q$ weakly, then $(X_i^N)_{1\leq i\leq N}$ is Q-chaotic.*

Note that $\lim_{N\to\infty} \mathcal{L}(\mu^N) = \delta_Q$ if and only if $\lim_{N\to\infty} \mu^N = Q$ in law or in probability, which constitutes a **law of large numbers**. Results on laws on path space are much stronger than results on the marginal processes and often imply them as below, see also Theorem 4.6 and Corollary 4.7 in Graham and Méléard [**5**].

THEOREM 6.2. *Let* $\lim_{N\to\infty} \mathcal{L}(\mu^N) = \delta_Q$ *weakly in* $\mathcal{P}(\mathcal{P}(\mathbb{D}_\mathcal{S}))$. *If moreover*

$$\forall T \geq 0\,, \quad \lim_{\varepsilon\to 0} \sup_{0\leq t\leq T} E_Q\Big(\sup_{t-\varepsilon<s<t+\varepsilon} |X_s - X_{s-}| \wedge 1\Big) = 0\,, \tag{6.4}$$

then $\lim_{N\to\infty} \bar{X}^N = (Q_t)_{t\geq 0}$ *in probability, for uniform convergence on bounded intervals on* $\mathbb{D}(\mathbb{R}_+, \mathcal{P}(\mathcal{S}))$, *and* $(Q_t)_{t\geq 0}$ *is continuous.*

6.2. Nonlinear martingale problems and Kolmogorov equations. The limit object is an inhomogeneous Markov law, which is nonlinear in the sense that the generator is a function of the marginal at time t of the law. Notably, the Kolmogorov equation is nonlinear. An appropriate tool for dealing with this is the nonlinear martingale problem on the canonical space.

For q in $\mathcal{P}(\mathcal{S})$, let $\mathcal{A}(q) : \phi \in L^\infty(\mathcal{S}) \mapsto \mathcal{A}(q)\phi \in L^\infty(\mathcal{S})$ be given for some uniformly bounded positive measure kernel J on $\mathcal{P}(\mathcal{S}) \times \mathcal{S}$ by

$$\forall \phi \in L^\infty(\mathcal{S})\,, \quad (\mathcal{A}(q)\phi)(x) = \int_\mathcal{S} \{\phi(y) - \phi(x)\} J(q,x,dy)\,. \tag{6.5}$$

DEFINITION 6.3. Let Q be in $\mathcal{P}(\mathbb{D}_\mathcal{S})$, $Q_s = \mathcal{L}(X_s)$ its marginal at time $s \geq 0$,

$$\phi(X_t) - \phi(X_0) - \int_0^t (\mathcal{A}(Q_s)\phi)(X_s)\,ds = M_t^\phi\,, \qquad t \geq 0\,,\ \phi \in L^\infty(\mathcal{S})\,. \tag{6.6}$$

Then Q solves the **nonlinear martingale problem** starting at q iff $Q_0 = q$ and $(M_t^\phi)_{t\geq 0}$ is a martingale under Q for the canonical filtration.

DEFINITION 6.4. Given a process $(R_t)_{t\geq 0}$ in $\mathcal{P}(\mathcal{S})$, Q in $\mathcal{P}(\mathbb{D}_\mathcal{S})$ solves the **martingale problem linearized at** $(R_t)_{t\geq 0}$ iff for all $\phi \in L^\infty(\mathcal{S})$, $\phi(X_t) - \phi(X_0) - \int_0^t (\mathcal{A}(R_s)\phi)(X_s)\,ds$, $t \geq 0$, is a martingale under Q.

REMARK 6.5. If the solution Q is unique, Q is an inhomogeneous Markov law with generators $\mathcal{A}(R_t)$, $t \geq 0$. If Q solves the martingale problem linearized at $(R_t)_{t\geq 0}$ and $(Q_t)_{t\geq 0} = (R_t)_{t\geq 0}$, then Q solves the nonlinear martingale problem.

If Q solves a martingale problem, then $(Q_t)_{t\geq 0}$ solves the evolution equation in weak form derived by taking the expectation of the martingale problem.

DEFINITION 6.6. The (deterministic) process $(Q_t)_{t\geq 0}$ taking values in $\mathcal{P}(\mathcal{S})$ solves the **nonlinear Kolmogorov equation** starting at q iff $Q_0 = q$ and

$$\langle \phi, Q_t \rangle = \langle \phi, Q_0 \rangle + \int_0^t \langle \mathcal{A}(Q_s)\phi, Q_s \rangle\,ds\,, \qquad t \geq 0\,,\ \phi \in L^\infty(\mathcal{S})\,. \tag{6.7}$$

This can be interpreted as $\partial_t Q_t(dy) = \langle J(Q_t,x,dy), Q_t(dx)\rangle - |J(Q_t,y)| Q_t(dy)$ and corresponds for discrete $\mathcal{S}$ to an infinite system of scalar differential equations.

DEFINITION 6.7. $(Q_t)_{t\geq 0}$ solves the **Kolmogorov equation linearized at** $(R_t)_{t\geq 0}$ iff for all $\phi \in L^\infty(\mathcal{S})$ and $t \geq 0$, $\langle \phi, Q_t \rangle = \langle \phi, Q_0 \rangle + \int_0^t \langle \mathcal{A}(R_s)\phi, Q_s \rangle\,ds$.

The nonlinear Kolmogorov equation can be considered as a fixed-point problem for $(R_t)_{t\geq 0} \mapsto (Q_t)_{t\geq 0}$. In this perspective, the following general result is an extension of Shiga and Tanaka [**7**] Lemma 2.3, and we refer to Graham, [**2**] Proposition 9.4.1 or [**3**] Proposition 2.3, for the proof.

THEOREM 6.8. *Assume that for some constants $B > 0$ and $K > 0$,*

$$|J(q,x)| \le B \quad \text{and} \quad |J(p,x) - J(q,x)| \le K|p-q|\,, \quad \forall p,q \in \mathcal{P}(\mathcal{S})\,,\ \forall x \in \mathcal{S}\,.$$

Then for any q in $\mathcal{P}(\mathcal{S})$, there is a unique solution Q in $\mathcal{P}(\mathbb{D}_\mathcal{S})$ for the nonlinear martingale problem (6.6) starting at q, and the marginal process $(Q_t)_{t\ge 0}$ is in $C^1(\mathbb{R}_+, \mathcal{P}(\mathcal{S}))$ and is the unique solution for the nonlinear Kolmogorov equation (6.7) starting at q. The solutions depend continuously on q.

References

[1] Ethier, S. and Kurtz, T.: *Markov processes*. New-York: John Wiley & Sons

[2] Graham, C.: Kinetic limits for large communication networks. In: *Modelling in Applied Sciences: A Kinetic Theory Approach*, Bellomo and Pulvirenti eds., Boston: Birkhauser 1999.

[3] Graham, C.: Chaoticity on path space for a queuing network with selection of the shortest queue among several. *Journal of Applied Probability* 37 (2000)

[4] Graham, C. and Méléard, S.: Chaos hypothesis for a system interacting through shared resources. *Probability Theory and Related Fields* 100, 157–173 (1994)

[5] Graham, C. and Méléard, S.: Stochastic particle approximations for generalized Boltzmann models and convergence estimates. *Annals of Probability* 28, 115–132 (1997)

[6] Martin, J.B. and Suhov, Yu.M.: Fast Jackson networks. Annals of Applied Probability 9, 854–870 (1999)

[7] Shiga, T. and Tanaka, H.: Central limit theorem for a system of Markovian particles with mean field interaction. *Z. Wahrscheinlichkeitstheorie verw. Gebeite* 69, 439–459 (1985)

[8] Sznitman, A.: Propagation of chaos. In: *École d'été de Saint-Flour 1989* (France). Lecture Notes in Mathematics 1464, 165–251. NewYork: Springer-Verlag

[9] Turner, S: The effect of increasing routing choice on resource pooling. *Probability in the Engineering and Informational Sciences* 12, 109–124 (1998)

[10] Vvedenskaya, N., Dobrushin, R. and Karpelevich, F.: Queuing system with selection of the shortest of two queues: an asymptotic approach. *Problems of Information Transmission* 32, 15–27 (1996)

[11] Whitt, W.: Blocking when service is required from several facilities simultaneously. *AT&T Technical Journal* 64, 1807–1856 (1985)

CMAP, ÉCOLE POLYTECHNIQUE, 91128 PALAISEAU, FRANCE.
E-mail address: carl@cmapx.polytechynique.fr

Contemporary Mathematics
Volume **275**, 2001

Maintenance of deteriorating machines with probabilistic monitoring and silent failures

Julie Simmons Ivy and Stephen M. Pollock

ABSTRACT. We characterize the structure of optimal policies for maintenance and replacement actions over a finite horizon. The context is machine monitoring when:

1. the machine can operate in either *good*, *worn* or *failed* states;
2. observations are related probabilistically to the state of the process; and
3. the machine's state is known with certainty *only* immediately after a replacement.

The last assumption, consistent with "silent" failures, distinguishes our results from others. We prove [using the theory and results of partially observable Markov decision processes (POMDPs)] that the policy that minimizes the total expected cost of system maintenance has a monotonic structure. This allows us to represent the optimal policy by a collection of decision rules characterized by at most three functions.

1. Introduction

Systems used in manufacturing naturally tend to wear, deteriorate, or break. Consequently, they must be maintained or replaced. Maintenance actions such as inspections, repairs, or replacements are taken to avoid operating the system in an undesirable state. Frequent inspections, repairs, or replacements may interrupt production and increase downtime and total maintenance costs. On the other hand, operating a system after it has deteriorated may also be very expensive. The decision maker attempts to postpone the system's failure through cost-effective maintenance, replacing (or repairing) the system only when necessary. To do so, the decision maker needs to be able to detect and correct any change in the quality of the system (i.e., its ability to operate), especially its failure, as soon as possible.

We analyze maintenance policies for a multi-state deteriorating system in which imperfect observations can be made. We use: information provided by the signals machines generate (in the form of sound, light, pressure, temperature, etc.); knowledge of the state transition distributions; and maintenance and operating costs to characterize the general structure of the maintenance policy which minimizes total

1991 *Mathematics Subject Classification.* Primary 90C40,90B25.

Key words and phrases. Partially Observable Markov Decision Processes (POMDP), Maintenance Models, Threshold Rules.

expected cost. We assume that all failures are "silent", i.e., that unless some action is taken, the system continues to operate under the *failed* condition and there is a cost, c_1, associated with such operation. We define the structure of the optimal policy for a three state system. We prove that, for a three-state model, the policy that minimizes the total expected cost has a "marginally monotonic" structure. This allows us to represent the optimal policy as a collection of decision rules characterized by (at most) three functions. The resulting optimal policy structures for Bernoulli and binomial monitoring distributions are computed numerically.

In the next section, we present background and preliminary results and review some known results. We present the model formulation in Section 3. We introduce and prove marginal monotonicity for finite horizon in Section 4. We present numerical examples in Section 5. We conclude in Section 6.

2. Models of Multi-State Deteriorating Systems

The concepts of state and state transition are used to model the multi-state deteriorating system. The system "state" is the condition of the system. A "state transition" is a change in the system condition. The concept of state allows one to focus on the system's features essential to the problem at hand, while the concept of state transition provides the mechanism for structuring the system's dynamic behavior (17). Most situations contain an element of uncertainty in the transitions of the system from one state to another.

In situations where this uncertainty depends only on the current state information, a Markov decision process (MDP) is a natural model of the system. An MDP is an optimization model for discrete-stage, stochastic sequential decision making (26). Decisions are made at points of time referred to as decision epochs. Let T denote the set of decision epochs, $T \equiv \{1, 2, \ldots, M\}$, where M is called the "horizon" of the problem. We assume decisions are made at the beginning of each decision epoch.

Many situations exist in which the decision maker does not know the current state of the system. For example, the state of a machine might include the condition of various internal and unobservable components of the machine, sensors used to measure the state may give noise-corrupted readings, and/or exact state observations may be costly. A partially observable Markov decision process (POMDP) is a generalization of a MDP that allows for gathering incomplete information regarding the state of the system. (See (10), (13), (11), (20) and (26) for overviews of theoretical and computational results, applications, and several generalizations of partially observable Markov decision processes and the standard MDP problem formulation.) Our objective is to use the POMDP structure to develop a method for selecting maintenance actions that minimize the total expected cost for operating a system. This cost includes the cost per unit time for operating the system in each state as well as the cost of performing actions to change the system state. Specifically, we determine the structure of the policy that minimizes the total expected cost of maintenance for a system whose states represent three conditions.

We define *maintenance* to be any repair or renewal action performed on the system. A maintenance policy is a set of rules, or a procedure, that prescribes the action to take, after observing available information: it is a function whose range is A and whose domain at time t is the vector $[x_1, x_2, \ldots, x_t]$ (the information provided by monitoring). The decision maker's objective is to select a policy that minimizes the system's total expected cost over a finite number of observations or the expected cost per unit time over an infinite number of observations.

Most of the multi-state models presented in the literature either assume that perfect information is available about the state of the system through monitoring or inspection (refer to (9) and (14))), or they assume that performing maintenance actions on a system with imperfect information will reveal the true state of the system (e.g., the system *renews* following a maintenance action, refer to (4), (22), (23), (24), (25) and (21)). The model for a system with N-levels of deterioration and perfect information was originally introduced by Derman (2) [extended by Klein (6)]. For this model, the optimal replacement rule is shown to be a "control limit" rule that is a function of the system state. This key result reduces the minimum size of the set of possible rules. Hopp and Wu (5) show that a simple "constant maintenance control limit" structure results if reward functions are separable and either the effect of maintenance actions is assumed to be independent of the current state or maintenance actions are assumed to be deterministic. Our model extends this work to include both imperfect monitoring and replacement and *state-dependent* repairs.

Imperfect maintenance refers to deterministic maintenance actions whose effect *depends* on the current state of the system. Therefore, since states are probabilistically known, their effect on the state of the system is also probabilistically known. This type of problem is the focus of the research presented in this paper. Imperfect maintenance models have been studied extensively for systems with two states.

Shiryaev (15) proved the Fixed Probability Threshold Rule for the two-state model with geometric failure times, i.e., that a "probability threshold rule" minimizes average cost per unit time. This rule is

> if π, the posterior probability of the system operating in the *good* condition, is less than or equal to some threshold value, then a checking action (a_1) should be taken.

For similar proofs, see (18) and (19). Albright (1) presents sufficient conditions for the optimal policy of partially observed systems with two states to be monotone in the information vector. Wang (21) proves that, for a non-geometric failure time distribution, and arbitrary monitoring distributions, a conditional probability threshold rule (CPTR) is an optimal checking policy for a single replacement problem. Threshold policies are an important class of optimal maintenance policies because they make it possible for the optimal policy to be summarized by the single parameter.

Ohnishi et al. (12) prove that the optimal cost function and the optimal policy function have monotone properties with respect to a MLR ordering by introducing two conditions: "available" and "unavailable or failure", that are assumed to be known with certainty. For each of these conditions there are two possible actions available to the decision maker, this reduces their problem to a variant of the two-state/two-action model. Gong et al. (3) analyze a problem similar to the one presented in this paper. However, they assume a threshold policy and focus on the methods for solving the problem of determining the optimal threshold value for minimizing total cost.

Lovejoy (8) presents a model that is most similar to the one presented in this paper. He examines discrete-time, finite, partially observed Markov decision processes and provides sufficient conditions for the optimal value function to be monotone on

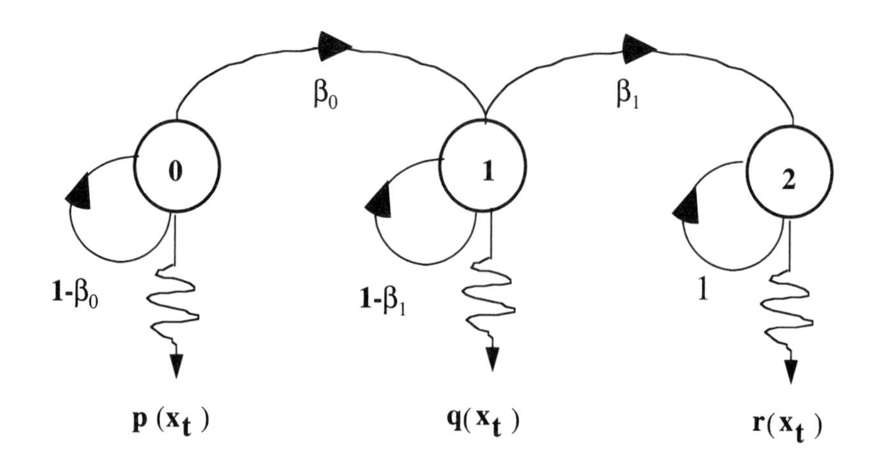

FIGURE 1. State Transition Diagram for Three-State Model

the space of state probability vectors ordered by likelihood ratios. However, as discussed in (16), the partial order presented in this paper is different from Lovejoy's and provides sufficient conditions for the expected cost function to be monotone in the state occupancy probability vector.

3. Three-State Model

3.1. State Transitions and Observations. We extend the two-state model by adding an intermediate state between the *good* and *bad* operating conditions. Consider a system that can be in one of three states, 0, 1, or 2 where 0 is *good*, 1 is *worn* (i.e., the system is still producing usable parts but is not in perfect working order), and 2 is *bad* or *failed*. Let $\mathcal{S} \equiv \{0,1,2\}$ be the set of possible system conditions, and $S_t \epsilon \mathcal{S}$ be the condition of the system at time t, $t = 1, 2, \ldots$.

The system starts in 0 and enters 1 after some operating time I_0, with realizations, $i_0 = 1, 2, \ldots$. After an additional operating time, I_1, with realizations, $i_1 = 1, 2, \ldots$, the system leaves state 1 and enters state 2. I_j is a discrete random variable with p.m.f., $f_j(i_j) = \beta_j(1-\beta_j)^{i_j-1}$, $j = 0, 1$. Once the system enters state 2, it remains there until some exogenous action is taken. Again, the "silent" failure aspect that is crucial to the discussion. The system's movement from states *good* to *worn* and finally to *bad* is illustrated in Fig. 1.

As in the two-state model, we assume monitoring is available. At every time, t, $t = 1, 2, 3, \ldots$, a random variable X_t is observed at no cost. The X_t are i.i.d. with p.d.f., $f_{X_t}(x)$. The relationship between the system state and the p.d.f of X_t is

$$f_{X_t}(x) = \begin{cases} p(x) & \text{if } S_t = 0 \\ q(x) & \text{if } S_t = 1 \\ r(x) & \text{if } S_t = 2 \end{cases} \tag{1}$$

where $p(x)$,$q(x)$, and $r(x)$ are *known* p.d.f.'s on x. The information provided by each observation, X_t, can be used to update the state occupancy probabilities.

TABLE 1. Summary of Effects of Actions on State.

Values in the Table represent the state that results from the action.

	Current State		
Action	0	1	2
a_0	0	1	2
a_1	0	0	1
a_2	0	0	0

Define the vector of observations through time t as $\underline{X}_t \equiv [X_1, X_2, \ldots, X_t]$ with the realization $\underline{x}_t \equiv [x_1, x_2, \ldots, x_t]$.

The state occupancy probabilities are the probabilities of the system operating in states 0, 1, and 2, respectively at time t, specifically,

$$\begin{aligned} \pi(t) &= Prob\{S_t = 0\} \\ \phi(t) &= Prob\{S_t = 1\} \\ \rho(t) &= Prob\{S_t = 2\} = 1 - \pi(t) - \phi(t) \end{aligned} \tag{2}$$

When the state occupancy probabilities are written without the index t, they refer to the current state probabilities.

3.2. Action Space. Three actions, $\{a_0, a_1, a_2\}$, are available to the decision maker for regulating the behavior of the system for maintenance.

a_0=: *Do Nothing*
a_1=: *Repair*
a_2=: *Renew*

The addition of the *repair* action is one of the critical differences between the three-state and two-state models. Unlike the *do nothing* and *renew* actions, the effect of the *repair* action on the system depends on the current state. Moreover, taking action a_1 does not provide perfect information about the next state (in contrast to (25), (4), (12), etc.). Table 1 summarizes the state that results from taking each of the actions, as a function of the current system condition.

Possible motivations for selecting each action are discussed below.

Do Nothing (a_0):: If the decision maker selects action a_0, he/she believes that the system is operating in the *good* state (0). Alternatively, the decision maker may select action a_0, even if he/she believes the system is in the *worn* state (1) because the system still produces usable parts and the penalty charged is moderate. In particular, it may not be necessary to repair the system if

1. it does not appear that the system will fail before the next observation;
2. it is not cost-effective to perform maintenance (e.g., it is not a good time to stop the machine or the resources are not available to perform maintenance).

Repair (a_1):: If the decision maker repairs the system while it is in state 1, the system returns to state 0. If the decision maker decides to *repair* while the system is in state 2, the system will be put into state 1. The decision to *repair* may be chosen because it may not be possible to *renew* at the current time due to resource and/or time constraints.

Renew (a_2):: The decision maker replaces the system (or performs *major* maintenance) to return the system to state 0. The system is renewed. Note that it is only after renewal that the state of the system is known.

3.3. Cost Structure. The cost structure is the same as for the two-state model with an additional operating cost per unit time associated with the *worn* state and a fixed cost associated with the *repair* action.

Define

$c_k \equiv$ cost per unit time of the system operating in state k, where $k = 0, 1, 2$

$$d_i \equiv \text{fixed cost of performing action } a_i,\ i = 0, 1, 2$$

Using the above definitions,

$$\begin{aligned} c\,(a_i|S = k) &\equiv \text{cost of action } a_i \text{ given current state } k \\ &\equiv d_i + c_{\max(k-i,0)} \end{aligned} \tag{3}$$

The expected cost of action a_k given state occupancy probabilities, π, ϕ, and ρ, is

$$L\,(a_0|\pi, \phi) = c_1\phi + c_2\rho$$

$$L\,(a_1|\pi, \phi) = d_1 + c_1\rho$$

$$L\,(a_2|\pi, \phi) = d_2$$

The dynamic program for the three-state model is similar to the two-state model, with the minimum expected total cost a function of both π and ϕ. Let

$C_k^*(\pi, \phi)$ =: Minimum expected cost when there are k observations remaining and probabilities π, ϕ (and $1 - \pi - \phi$) of being in states 0, 1 (and 2), respectively.

Then

$$C_k^*(\pi, \phi) = \min_{l \in (0,1,2)} \left\{ L\,(a_l|\pi, \phi) + E_X\left[C_{k-1}^*\left(\pi'(X|a_l), \phi'(X|a_l)\right)\right]\right\} \tag{4}$$

$$k = 1, 2, \ldots, M$$

where

$$C_0^*(\pi, \phi) = 0$$

$$E_X\left[C_{k-1}^*\left(\pi'(X|a_l), \phi'(X|a_l)\right)\right] = \int_x C_{k-1}^*\left(\pi'(x|a_l), \phi'(x|a_l)\right) f_X(x)dx$$

$$f_X(x) = \pi'(x|a_l)p(x) + \phi'(x|a_l)q(x) + \rho'(x|a_l)r(x),$$

$$\begin{aligned} \pi'(x|a_l) &= \text{probability } S = 0 \text{ given } x \text{ and action } a_l \\ \phi'(x|a_l) &= \text{probability } S = 1 \text{ given } x \text{ and action } a_l \\ \rho'(x|a_l) &\equiv 1 - \pi'(x|a_l) - \phi'(x|a_l) \end{aligned}$$

Using Bayes' Rule, it can be shown that

$$\begin{aligned} \pi'(x|a_0) &= \pi' = \frac{(1-\beta_0)\pi p(x_m)}{\pi p(x_m) + \phi q(x_m) + \rho r(x_m)} \\ \phi'(x|a_0) &= \phi' = \frac{(1-\beta_1)\phi q(x_m) + \beta_0 \pi p(x_m)}{\pi p(x_m) + \phi q(x_m) + \rho r(x_m)} \end{aligned} \tag{5}$$

(6)
$$\begin{aligned}\pi'(x|a_1) &= \pi' + \phi' \\ \phi'(x|a_1) &= 1 - \pi' - \phi'\end{aligned}$$

(7)
$$\begin{aligned}\pi'(x|a_2) &= 1 \\ \phi'(x|a_2) &= 0\end{aligned}$$

The solution, $C_M^*(\pi, \phi)$, yields the optimal policy for the M-period finite horizon three-state model.

For notational convenience, the dynamic program of Eq.(4) can be rewritten:

(8)
$$C_k^*(\pi, \phi) = \min \begin{cases} \Omega_k(\pi, \phi) \\ \Psi_k(\pi, \phi) \\ Z_k \end{cases}$$

The first term in Eq.(8) is the expected cost of the *Do Nothing* action:

(9)
$$\Omega_k(\pi, \phi) \equiv c_1\phi + c_2(1 - \pi - \phi) + E_X\left[C_{k-1}^*(\pi'(X|\Omega, \pi, \phi), \phi'(X|\Omega, \pi, \phi))\right]$$

where

$$\pi'(x|\Omega, \pi, \phi) \equiv \pi'(x|a_0)$$

and

$$\phi'(x|\Omega, \pi, \phi) \equiv \phi'(x|a_0)$$

are the updated probabilities of the system operating in the states 0 and 1 given the *do nothing* action, a_0, is taken and x has been observed. The second term is the expected cost of the *Repair* action:

(10)
$$\Psi_k(\pi, \phi) \equiv d_1 + c_1\rho + E_X\left[C_{k-1}^*(\pi'(X|\Psi, \pi, \phi), \phi'(X|\Psi, \pi, \phi))\right]$$

where

$$\pi'(x|\Psi, \pi, \phi) \equiv \frac{p(x)\pi + (1 - \beta_1)q(x)\phi}{\pi p(x) + \phi q(x) + \rho r(x)} = \pi'(x|a_1)$$

and

$$\phi'(x|\Psi, \pi, \phi) \equiv \frac{\beta_1 q(x)\phi + \rho r(x)}{\pi p(x) + \phi q(x) + \rho r(x)} = \phi'(x|a_1)$$

are the updated probabilities of the system operating in the *good* and *worn* states given the *repair* action, a_1, is taken and x has been observed. Notice above $\pi'(x|\Psi, \pi, \phi) + \phi'(x|\Psi, \pi, \phi) = 1$. The third term, the expected cost of the *Renew* action, a_2, is constant in π and ϕ:

(11)
$$Z_k \equiv d_2 + C_{k-1}^*(1, 0)$$

4. Marginal Monotonicity

Recall for the two-state model, the optimal decision rule is a probability threshold rule characterized by a threshold π^*. We prove that a similar decision structure exists for the three-state model. As shown in Fig. 2, for the three-state model, the collection of possible (π, ϕ) points is represented by the simplex

$$\Delta = \{(\pi, \phi) : 0 \leq \pi \leq 1, 0 \leq \phi \leq 1, \pi + \phi \leq 1\}.$$

For each (π, ϕ) pair in the simplex, there is an action (a^*) that minimizes total expected cost. We now prove that the three-state optimal policy has a "marginally" monotonic structure, also characterized by a threshold-type of rule.

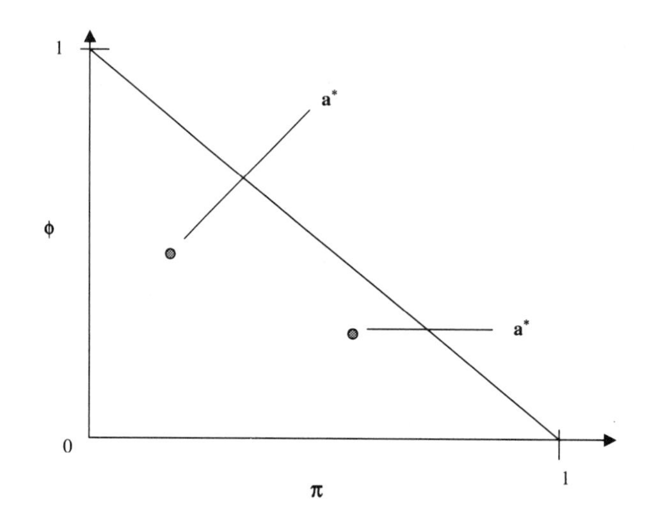

FIGURE 2. Decision Simplex for the Three-State Model.

DEFINITION 4.1. *A **decision region**, $R_i(\pi, \phi) \in \Re^2$, is the set of $(\pi, \phi) \in \Delta$ for which $a^* = a_i$ for $i \in 0, 1, 2$.*

DEFINITION 4.2. *A decision region, $R_i(\pi, \phi) \in \Re^2$, is **marginally monotonic** if*

$R_i(\pi, \phi^o)$: is convex in π $\forall$ ϕ^o and
$R_i(\pi^o, \phi)$: is convex in ϕ $\forall$ π^o.

That is, a region, $R_i(\pi, \phi)$, is marginally monotonic if any one dimensional projection of the region in π (or ϕ) is convex in π (or ϕ).

Given a one-dimensional projection in π, i.e., on the line $\phi = \phi^o$, if the decision regions are marginally monotonic then the decision rule has the following form: There exist $\pi_1^*(\phi^o)$ and $\pi_2^*(\phi^o)$ for which

(i): it is optimal to *renew* if $\pi \leq \pi_1^*(\phi^o)$.
(ii): it is optimal to *repair* if $\pi_1^*(\phi^o) \leq \pi \leq \pi_2^*(\phi^o)$, and
(iii): it is optimal to *do nothing* if $\pi_2^*(\phi^o) \leq \pi$.

Given a one-dimensional projection on the line $\pi = \pi^o$, the optimal decision rule has the same type of convex structure. Figure 3 illustrates this structure, where $\phi^o = 0.16$, $\pi_1^*(\phi^o) \approx 0.45$, $\pi_2^*(\phi^o) \approx 0.76$.

We now show our major result, a threshold-like rule for the three-state model:

THEOREM 4.1.

$$C_k^*(\pi, \phi) = \begin{cases} \Omega_k(\pi, \phi) & (\pi, \phi) \epsilon R_1^k \\ \Psi_k(\pi, \phi) & (\pi, \phi) \epsilon R_2^k \\ Z_k & (\pi, \phi) \epsilon R_3^k \end{cases}$$

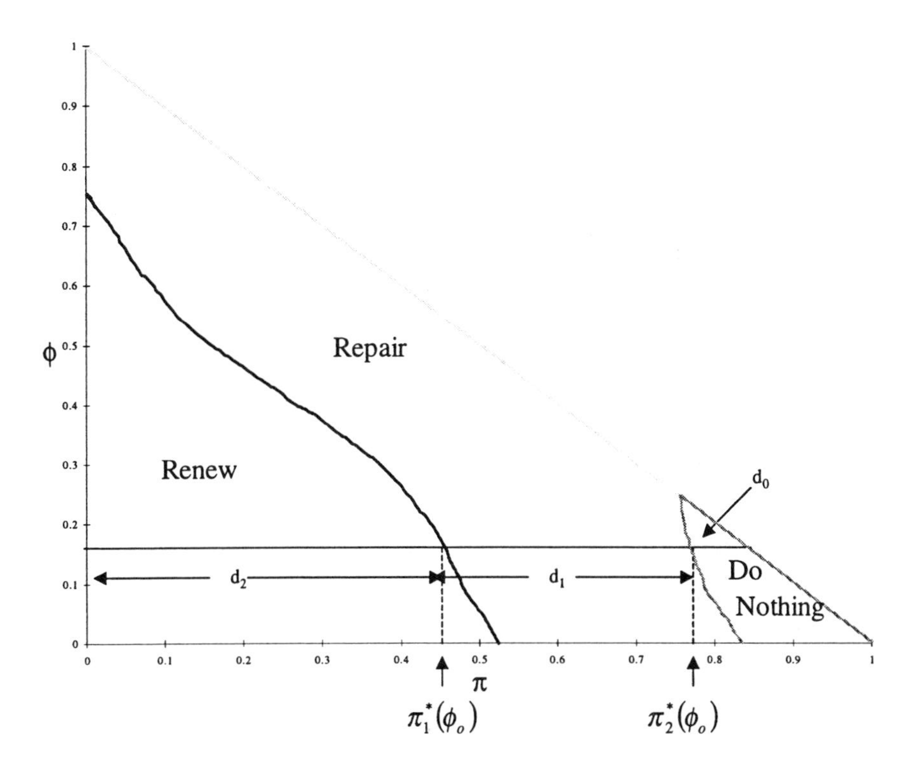

FIGURE 3. Decision Regions with respect to π and ϕ

where $R_i^k \equiv R_i(\pi, \phi|k)$ are marginally monotonic decision regions for all k.

COROLLARY 4.1. *For $k = 0, 1, \ldots$, the action that minimizes the expected total cost is*

$$a_k^*(\pi, \phi) = \begin{cases} a_0 & (\pi, \phi)\epsilon R_1^k \\ a_1 & (\pi, \phi)\epsilon R_2^k \\ a_2 & (\pi, \phi)\epsilon R_3^k \end{cases}$$

The proof of Theorem 4.1 is based on the following argument: From Eq.(11), the cost to *renew*, Z_k, is constant with respect to π and ϕ. If $\Omega_k(\pi, \phi)$, $\Psi_k(\pi, \phi)$ and $\Omega_k(\pi, \phi) - \Psi_k(\pi, \phi)$ are monotone functions of both π and ϕ, then each pair of $\{\Omega_k(\pi, \phi^o), \Psi_k(\pi, \phi^o), Z_k\}$ intercept each other for at most one π. The same holds for (π^o, ϕ). Conditions a, b and Lemma 4.1 ensure that each pair of $\{\Omega_k(\pi, \phi^o), \Psi_k(\pi, \phi^o)$, and $Z_k\}$ intercept each other for at most one π and the same holds for (π^o, ϕ).

Condition (a): For all $\{\phi : 0 \leq \phi \leq 1\}$

1.: **either:** $\frac{\partial \Omega_k(\pi,\phi)}{\partial \pi} \geq 0$,

or: $\frac{\partial \Omega_k(\pi,\phi)}{\partial \pi} \leq 0$.

and

2.: **either:** $\frac{\partial \Psi_k(\pi,\phi)}{\partial \pi} \geq 0$,
or: $\frac{\partial \Psi_k(\pi,\phi)}{\partial \pi} \leq 0$.

and

3.: **either:** $\frac{\partial(\Omega_k(\pi,\phi)-\Psi_k(\pi,\phi))}{\partial \pi} \geq 0$,
or: $\frac{\partial(\Omega_k(\pi,\phi)-\Psi_k(\pi,\phi))}{\partial \pi} \leq 0$.

Condition (b): For all $\{\pi : 0 \leq \pi \leq 1\}$

1.: **either:** $\frac{\partial \Omega_k(\pi,\phi)}{\partial \phi} \geq 0$,
or: $\frac{\partial \Omega_k(\pi,\phi)}{\partial \phi} \leq 0$.

and

2.: **either:** $\frac{\partial \Psi_k(\pi,\phi)}{\partial \phi} \geq 0$,
or: $\frac{\partial \Psi_k(\pi,\phi)}{\partial \phi} \leq 0$.

and

3.: **either:** $\frac{\partial(\Omega_k(\pi,\phi)-\Psi_k(\pi,\phi))}{\partial \phi} \geq 0$,
or: $\frac{\partial(\Omega_k(\pi,\phi)-\Psi_k(\pi,\phi))}{\partial \phi} \leq 0$.

To demonstrate conditions a and b and thus prove Theorem 4.1, we establish the following preliminary results.

LEMMA 4.1. *$C_k^*(\underline{\pi})$ is a piecewise-linear and convex function of $\underline{\pi} = [\pi(0), \pi(1), \pi(2)]$ and can be expressed as:*

$$C_k^*(\underline{\pi}) = \min_m \left[\sum_{i=0}^{N} \alpha_i^m(\dot{k})\pi(i) \right] \tag{12}$$

for some set of vectors $\underline{\alpha}^m(k) = [\alpha_0^m(k), \alpha_1^m(k), \ldots, \alpha_N^m(k)]$, $m = 1, 2, \ldots$.

PROOF. Follows from Smallwood and Sondik (17), details of the proof are given in Simmons (16). □

LEMMA 4.2. *For $k = 1$, conditions a and b hold.*

PROOF. Follows directly from the linearity of $\Omega_1(\pi, \phi)$, $\Psi_1(\pi, \phi)$, and Z_1. □

LEMMA 4.3. *$C_1^*(\pi, \phi)$ is a monotonic function of π.*

PROOF. Follows from Lemma 4.2. □

By Lemmas 4.1, 4.2, and 4.3, we have shown that a marginally monotonic policy holds for $k = 1$. Specifically,

For $\Omega_1(\pi, \phi) = \Psi_1(\pi, \phi)$: on the line $\pi = 1 - \phi + \frac{c_1\phi - d_1}{c_2 - c_1}$,
For $\Omega_1(\pi, \phi) = Z_1$: on the line $\pi = 1 - \phi + \frac{c_1\phi - d_2}{c_2}$, and
For $\Psi_1(\pi, \phi) = Z_1$: on the line $\pi = 1 - \phi + \frac{d_1 - d_2}{c_1}$

creating the optimal cost function:

$$C_1^*(\pi, \phi) = \begin{cases} \Omega_1(\pi, \phi) & \text{if } (\pi, \phi) \in R_1^1 \\ \Psi_1(\pi, \phi) & \text{if } (\pi, \phi) \in R_2^1 \\ Z_1 & \text{if } (\pi, \phi) \in R_3^1 \end{cases}$$

with associated decision regions:

$$a_1^*(\pi, \phi) = \begin{cases} a_0 & \text{if } (\pi, \phi) \in R_1^1 \\ a_1 & \text{if } (\pi, \phi) \in R_2^1 \\ a_2 & \text{if } (\pi, \phi) \in R_3^1 \end{cases}$$

where

$$R_1^1 = \left\{(\pi,\phi) : \pi + \left(1 - \frac{c_1}{c_2 - c_1}\right)\phi \geq 1 - \frac{d_1}{c_2 - c_1},\ \pi + \left(1 - \frac{c_1}{c_2}\right)\phi \geq \left(1 - \frac{d_2}{c_2}\right)\right\}$$

$$R_2^1 = \left\{(\pi,\phi) : \pi + \left(1 - \frac{c_1}{c_2 - c_1}\right)\phi \leq 1 - \frac{d_1}{c_2 - c_1},\ \pi + \phi \geq \left(1 + \frac{d_1 - d_2}{c_1}\right)\right\}$$

$$R_3^1 = \left\{(\pi,\phi) : \pi + \left(1 - \frac{c_1}{c_2}\right)\phi \leq \left(1 - \frac{d_2}{c_2}\right),\ \pi + \phi \leq \left(1 + \frac{d_1 - d_2}{c_1}\right)\right\}$$

LEMMA 4.4. *$C_1^*(\pi'(X|\Omega,\pi,\phi),\phi'(X|\Omega,\pi,\phi)) - C_1^*(\pi'(X|\Psi,\pi,\phi),\phi'(X|\Psi,\pi,\phi))$ is a monotonic function of π for all X.*

PROOF. Using the results from Lemmas 4.2 and 4.3 and the definitions of $\pi'(X|\Omega,\pi,\phi)$, $\phi'(X|\Omega,\pi,\phi)$, $\pi'(X|\Psi,\pi,\phi)$, and $\phi'(X|\Psi,\pi,\phi)$, we can show that the derivative of $C_1^*(\pi'(X|\Omega,\pi,\phi),\phi'(X|\Omega,\pi,\phi)) - C_1^*(\pi'(X|\Psi,\pi,\phi),\phi'(X|\Psi,\pi,\phi))$ with respect to π is greater than or equal to zero. □

LEMMA 4.5. *For $l \geq 1$, $C_{l+1}^*(\pi,\phi)$ is a monotonic function of π.*

PROOF. For some $l > 1$, assume $C_l^*(\pi,\phi)$ is a monotonic function of π and ϕ and, in particular, for any $0 \leq \phi \leq 1$,

1. $\frac{\partial C_l^*(\pi,\phi)}{\partial \pi} \geq 0$
2. $\frac{\partial \Omega_l^*(\pi,\phi)}{\partial \pi} \geq 0$ and
3. $\frac{\partial \Psi_l^*(\pi,\phi)}{\partial \pi} \geq 0$

Recall, from Eq.(8),

(13)
$$C_{l+1}^*(\pi,\phi) = \min \begin{cases} \Omega_{l+1}(\pi,\phi) = c_1\phi + c_2\rho + E\left[C_l^*\left(\pi'(X|\Omega,\pi,\phi),\phi'(X|\Omega,\pi,\phi)\right)\right] \\ \Psi_{l+1}(\pi,\phi) = d_1 + c_1\rho + E\left[C_l^*\left(\pi'(X|\Psi,\pi,\phi),\phi'(X|\Psi,\pi,\phi)\right)\right] \\ Z_{l+1} = d_2 + C_l^*(1,0) \end{cases}$$

1. Since the first terms in $\Omega_{l+1}(\pi,\phi)$ and $\Psi_{l+1}(\pi,\phi)$, i.e., $c_1\phi + c_2\rho$ and $d_1 + c_1\rho$, are monotone with respect to π and ϕ, we need to show that $E\left[C_l^*(\cdot,\cdot)\right]$ are also monotone with respect to π and ϕ. In order to do this we show that $\pi'(X|\Omega,\pi,\phi)$, $\phi'(X|\Omega,\pi,\phi)$, $\pi'(X|\Psi,\pi,\phi)$, and $\phi'(X|\Psi,\pi,\phi)$ are monotone with respect to π and ϕ for all X.

   (1.a): For all X it can be shown, after using straight forward algebra, that

   (i):

$$\frac{\partial \pi'(x|\Omega,\pi,\phi)}{\partial \pi} \geq 0 \tag{14}$$

   (ii):

$$\frac{\partial \phi'(x|\Omega,\pi,\phi)}{\partial \pi} \geq 0 \tag{15}$$

   (iii):

$$\frac{\partial \pi'(x|\Psi,\pi,\phi)}{\partial \pi} \geq 0 \tag{16}$$

(iv):

$$\frac{\partial \phi'(x|\Psi,\pi,\phi)}{\partial \pi} \geq 0 \tag{17}$$

(1.b): Since by assumption 1, $\frac{\partial C_l^*(\pi,\phi)}{\partial \pi} \geq 0$, using Eq.(14), gives

$$\frac{\partial C_l^*(\pi'(x|\Omega,\pi,\phi),\phi'(x|\Omega,\pi,\phi))}{\partial \pi} \geq 0 \tag{18}$$

for all X. Equation (18) implies that

$$\frac{\partial E_X\left[C_l^*(\pi'(X|\Omega,\pi,\phi),\phi'(X|\Omega,\pi,\phi))\right]}{\partial \pi} \geq 0 \tag{19}$$

2. By Lemma 4.2, $\Omega_1(\pi,\phi)$ is non-decreasing in π, i.e.,

$$\frac{\partial\left(c_1\phi + c_2\rho\right)}{\partial \pi} \geq 0.$$

It follows that

$$\frac{\partial \Omega_{l+1}(\pi,\phi)}{\partial \pi} \geq 0.$$

A similar argument to that leading to Eq.(17) shows that $\frac{\partial \pi'(x|\Psi,\pi,\phi)}{\partial \pi} \geq 0$ and

$$\frac{\partial C_l^*(\pi'(x|\Psi,\pi,\phi),\phi'(x|\Psi,\pi,\phi))}{\partial \pi} \geq 0 \tag{20}$$

for all X. Eq.(20) implies that

$$\frac{\partial E_X\left[C_l^*(\pi'(X|\Psi,\pi,\phi),\phi'(X|\Psi,\pi,\phi))\right]}{\partial \pi} \geq 0$$

3. Finally, since

$$\frac{\partial\left(d_1 + c_1\rho\right)}{\partial \pi} \geq 0$$

it follows from Assumption 3 that

$$\frac{\partial \Psi_{l+1}(\pi,\phi)}{\partial \pi} \geq 0$$

4. We can show, using a parallel argument, that

$$\frac{\partial \Omega_{k+1}(\pi,\phi)}{\partial \phi} \geq 0$$

and

$$\frac{\partial \Psi_{k+1}(\pi,\phi)}{\partial \phi} \geq 0$$

□

LEMMA 4.6. *For $l \geq 1$, $\Omega_{l+1}(\pi,\phi) - \Psi_{l+1}(\pi,\phi)$is a monotonic function of π.*

PROOF. For some $l > 1$, assume

$$C_l^*(\pi'(X|\Omega,\pi,\phi),\phi'(X|\Omega,\pi,\phi)) - C_l^*(\pi'(X|\Psi,\pi,\phi),\phi'(X|\Psi,\pi,\phi)) \tag{21}$$

is a monotonic function of π and ϕ for all X.

Recall, from Eq.(8),

(22)

$$C^*_{l+1}(\pi,\phi)=\min\begin{cases}\Omega_{l+1}(\pi,\phi)=c_1\phi+c_2\rho+E\left[C^*_l\left(\pi'(X|\Omega,\pi,\phi),\phi'(X|\Omega,\pi,\phi)\right)\right]\\ \Psi_{l+1}(\pi,\phi)=d_1+c_1\rho+E\left[C^*_l\left(\pi'(X|\Psi,\pi,\phi),\phi'(X|\Psi,\pi,\phi)\right)\right]\\ Z_{l+1}=d_2+C^*_l(1,0)\end{cases}$$

By Lemma 4.4 the first term in $\Omega_{l+1}(\pi,\phi)-\Psi_{l+1}(\pi,\phi)$, i.e., $c_1\phi+(c_2-c_1)\rho-d_1$ is a monotone function of π (and ϕ). And by induction using assumption 21, $E_X[C^*_l(\pi'(X|\Omega,\pi,\phi),\phi'(X|\Omega,\pi,\phi))]-E_X[C^*_l(\pi'(X|\Psi,\pi,\phi),\phi'(X|\Psi,\pi,\phi))]$ is also monotone function of π (and ϕ). □

PROOF. We now prove Theorem 4.1 by induction

1. For $k=1$:
   (i): By Lemma 3.1, $\Omega_1(\pi,\phi)$ and $\Psi_1(\pi,\phi)$ are monotonic functions of both π and ϕ,
   and:
   (ii): By Lemma 3.2, $C^*_1(\pi,\phi)$ is monotonic with respect to π and ϕ.
2. Assume for some $l\geq 1$:
   (i): $\Omega_l(\pi,\phi)$ and $\Psi_l(\pi,\phi)$ are monotonic functions of both π and ϕ,
   and:
   (ii): $C^*_l(\pi,\phi)$ is monotonic with respect to both π and ϕ.
3. By Lemma 3.3, $\Omega_{l+1}(\pi,\phi)$ and $\Psi_{l+1}(\pi,\phi)$ are monotonic functions of both π and ϕ

Thus the induction is complete.

Since $\Omega_k(\pi,\phi)$ and $\Psi_k(\pi,\phi)$ are monotonic functions of both π and ϕ, and Lemma 4.1 shows that $C^*_k(\pi,\phi)$ is piecewise-linear and convex in π and ϕ, it follows that the decision regions R^k_i, $i=1,2,3$ are marginally monotonic. □

Notice that this definition of monotonicity differs from both White's (25) and Lovejoy's (7). For example, if $\pi=\frac{1}{3}$, $\phi=\frac{1}{3}$ and $\pi'=0$, $\phi'=\frac{1}{3}$,

a): White's ordering, i.e., $\sum_{i=q}^{2}\pi(i)\geq\sum_{i=q}^{2}\pi'(i)$ for $q=0,1,2$ (where $\pi(0)=\pi$, $\pi(1)=\phi$ and $\pi(2)=1-\pi-\phi$), is **satisfied**.
b): Lovejoy's MLR ordering, i.e., if $i\geq i'$ then $\pi(i)\pi'(i')\geq\pi(i')\pi'(i)$, is **violated** for $i=2$ and $i'=1$.
c): Marginal monotonicity, i.e., $(\pi,\phi^o,1-\pi-\phi^o)\geq(\pi',\phi^o,1-\pi'-\phi^o)$ and $(\pi^o,\phi,1-\pi^o-\phi)\geq(\pi^o,\phi',1-\pi^o-\phi')$, is **satisfied**.

However, if $\pi=\frac{1}{3}$, $\phi=\frac{1}{3}$ and $\pi'=0$, $\phi'=0$,

a): White's (25) partial ordering is **satisfied.**
b): Lovejoy's (7) partial ordering are **satisfied.**
c): Marginal monotonicity is **violated** because more than one component is changed.

As illustrated by the above examples, the monotonicity required by our model is:

1. a more restrictive partial ordering than the first-order stochastic dominance used by White (25); and
2. a different partial ordering than the MLR by Lovejoy (7)

TABLE 2. Parameter Values for Base Case.

State	*Deterioration Rate*	*State Operating Cost*	*Action a_i Cost*	*Observation Distribution Parameter*
i	β_i	c_i, *per unit time*	d_i, *fixed*	
0	0.2	0	0	$a = 0.9$
1	0.2	1	0.75	$b = 0.6$
2		3	1.50	$u = 0.2$

in that it requires a componentwise partial ordering and monotonicity of the expected total cost function with respect to each component of the state space. Note also that the monotonicity discussed here is for *policies*, not for the value function.

5. Examples

We present examples of decision regions obtained by numerically solving the dynamic program presented in Section 2, with π and ϕ discretized on a lattice with increments of 0.01. Linear interpolation of $C_k^*(\pi, \phi)$ is used for values of π and ϕ between the lattice points.

5.1. Bernoulli Monitoring. Assume the monitoring observations have Bernoulli distributions:

$$p(x) = \begin{cases} a & \text{if } x = 1 = \text{non-defective} \\ 1 - a & \text{if } x = 0 = \text{defective} \end{cases}$$

$$q(x) = \begin{cases} b & \text{if } x = 1 = \text{non-defective} \\ 1 - b & \text{if } x = 0 = \text{defective} \end{cases}$$

and

$$r(x) = \begin{cases} u & \text{if } x = 1 = \text{non-defective} \\ 1 - u & \text{if } x = 0 = \text{defective} \end{cases}$$

where $0 \leq u \leq b \leq a \leq 1$. The Bernoulli distribution can reasonably represent a process in which a single observation is taken at each decision epoch.

5.1.1. *Base Case.* Figure 4 shows the optimal decision regions for $k = 50$ and the parameters given in Table 2. For the examples that follow, all parameters are those given in Table 2 except for the parameter being varied. The unsmooth nature of the boundaries of the decision regions in Fig. 4 (and the figures that follow) is due to the discretization of π and ϕ.

5.1.2. *Horizon.* As shown in Fig. 5, the general shape and size of the decision regions do not change significantly with additional increases in k. In fact, as implied in Section 3, as the horizon grows, the decision regions become "stationary", converging in π and ϕ. For the example shown in Fig. 5, the decision regions converge numerically, for $k \geq 5$. Figure 6 shows that the expected cost per unit time increases significantly for small k, but approaches a constant for larger values of k. For $k \geq 50$, the optimal policy and the cost per unit time are relatively insensitive to the horizon length.

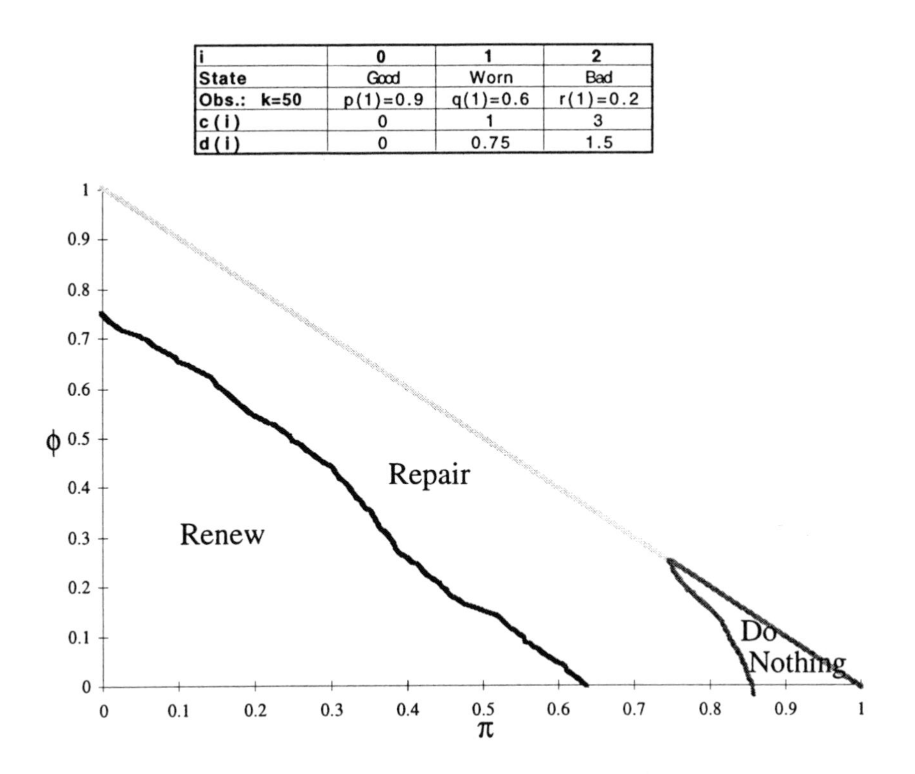

i	0	1	2
State	Good	Worn	Bad
Obs.: k=50	p(1)=0.9	q(1)=0.6	r(1)=0.2
c(i)	0	1	3
d(i)	0	0.75	1.5

FIGURE 4. Decision Regions for Base Case.

5.1.3. *Approximation to Optimal Policies.* As shown in Fig. 5, although the decision regions are marginally monotone, they are not necessarily convex. This significantly complicates formulating a simple "threshold" type of decision rule. However, the optimal decision region boundaries can be *approximated* by simple linear functions of π and ϕ (see Fig. 7).

To determine the effect of using such an approximate policy, we can compare the expected cost per unit time for each. For the example in Fig. 7, the expected cost per unit time increases by less than one percent, from 0.3789 to 0.3821. As shown in Table 3, similar approximations to the optimal policies of examples generated by varying the model parameters also increase the expected cost per unit time by less than one percent. These results suggest that simple linear decision rules do not significantly increase the cost per unit time.

5.1.4. *Comparison to Two State Case.* The three-state model is arguably more representative than a two-state model of a *true* deteriorating environment. Most systems and processes fail gradually, not abruptly, and an early stage of deterioration, if detected, can be used to make repairs.

We can compare the expected cost per unit time of a system well-described by the three-state/three-action model (see Fig. 8a) to that obtained by a policy developed assuming it is described by a two-state/two-action model (shown in Fig. 8b)

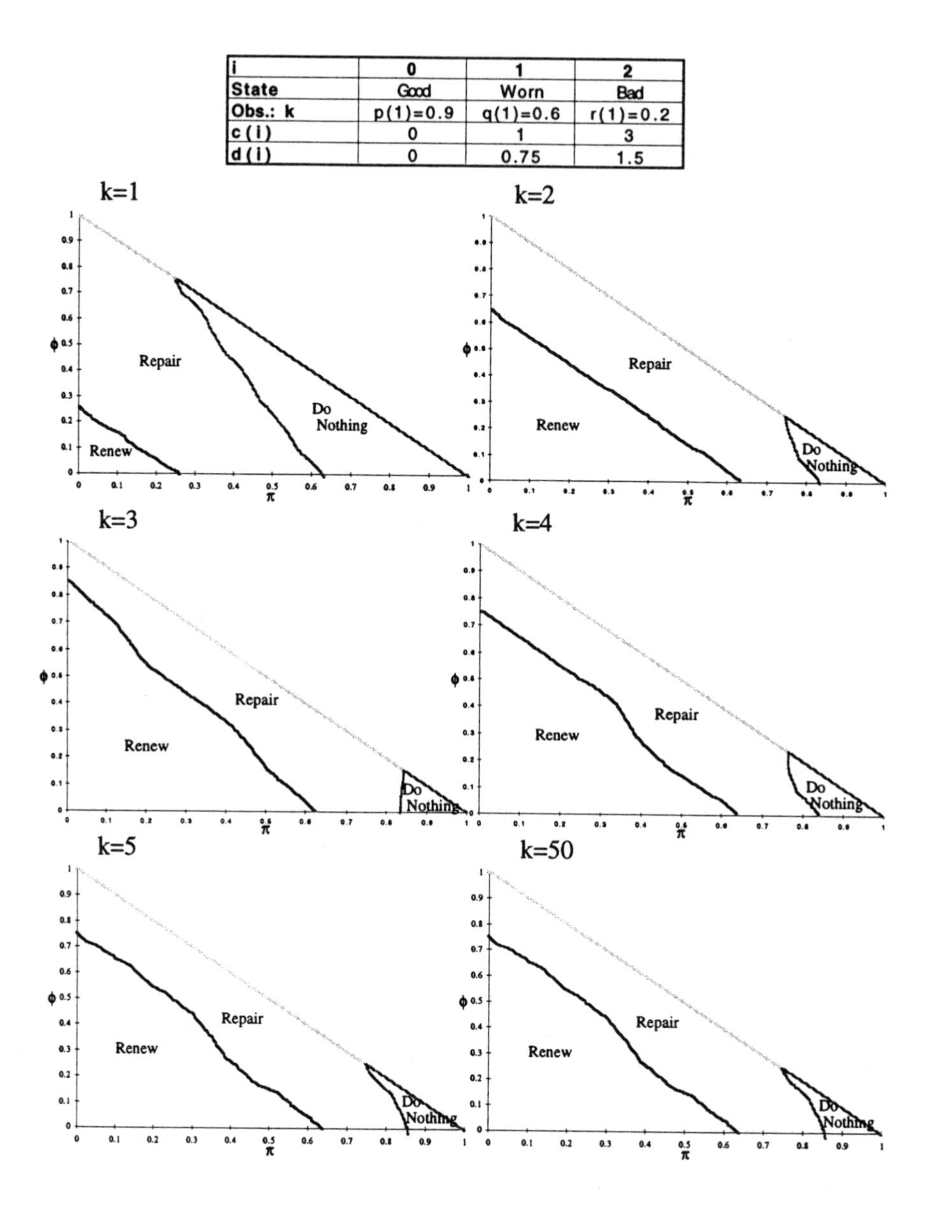

i	0	1	2
State	Good	Worn	Bad
Obs.: k	p(1)=0.9	q(1)=0.6	r(1)=0.2
c(i)	0	1	3
d(i)	0	0.75	1.5

FIGURE 5. Decision Regions as k increases.

with hazard, $\gamma = \frac{1}{\frac{1}{\beta_0}+\frac{1}{\beta_1}}$. This demonstrates the effect of including the third (*worn*) state (and the additional, *repair* action.)

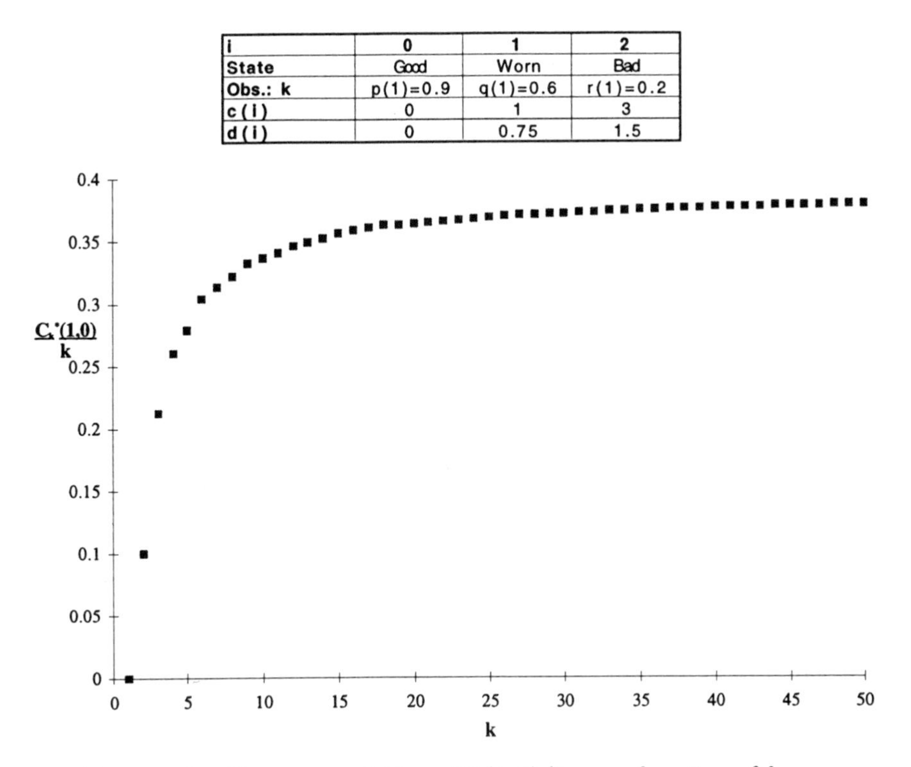

i	0	1	2
State	Good	Worn	Bad
Obs.: k	p(1)=0.9	q(1)=0.6	r(1)=0.2
c(i)	0	1	3
d(i)	0	0.75	1.5

FIGURE 6. Cost per unit time, $C_k^*(1,0)/k$, as a function of k.

Table 4 shows that the decision maker pays for the simplicity of the two-state model with a significant increase in the expected cost per unit time. The maintenance policy for the two-state model does not have the *repair* action and lacks the information provided by the *worn* state. The inclusion of this state and action results in an approximately 20% reduction in the expected cost per unit time for $0.10 \leq \gamma \leq 0.17$ with $\beta_0 = 0.2$ and for $0.13 \leq \gamma \leq 0.29$ with $\beta_0 = 0.4$. As γ increases with each $\beta_0 > 0.4$, the reduction in expected cost per unit time decreases for larger γ with $\beta_0 = 0.8$ and $\beta_0 = 0.6$ until it reaches 0.015 for $\gamma = 0.5$ and $\beta_0 = 1.0$ (in this case $\beta_0 = \beta_1 = 1$ and for these three-state hazards, the optimal three-state solution is to **never** *repair*).

5.2. Binomial Monitoring. In Section 4.1, we assumed that the observations were Bernoulli distributed; here, we assume that the observations are binomially distributed. Specifically, we define the distributions for the binomial observations as follows:

$$p(x){:} \; = \text{Bin}(n,a) = \binom{n}{x} a^x (1-a)^{n-x}$$

$$q(x){:} \; = \text{Bin}(n,b) = \binom{n}{x} b^x (1-b)^{n-x}$$

$$r(x){:} \; = \text{Bin}(n,u) = \binom{n}{x} u^x (1-u)^{n-x}$$

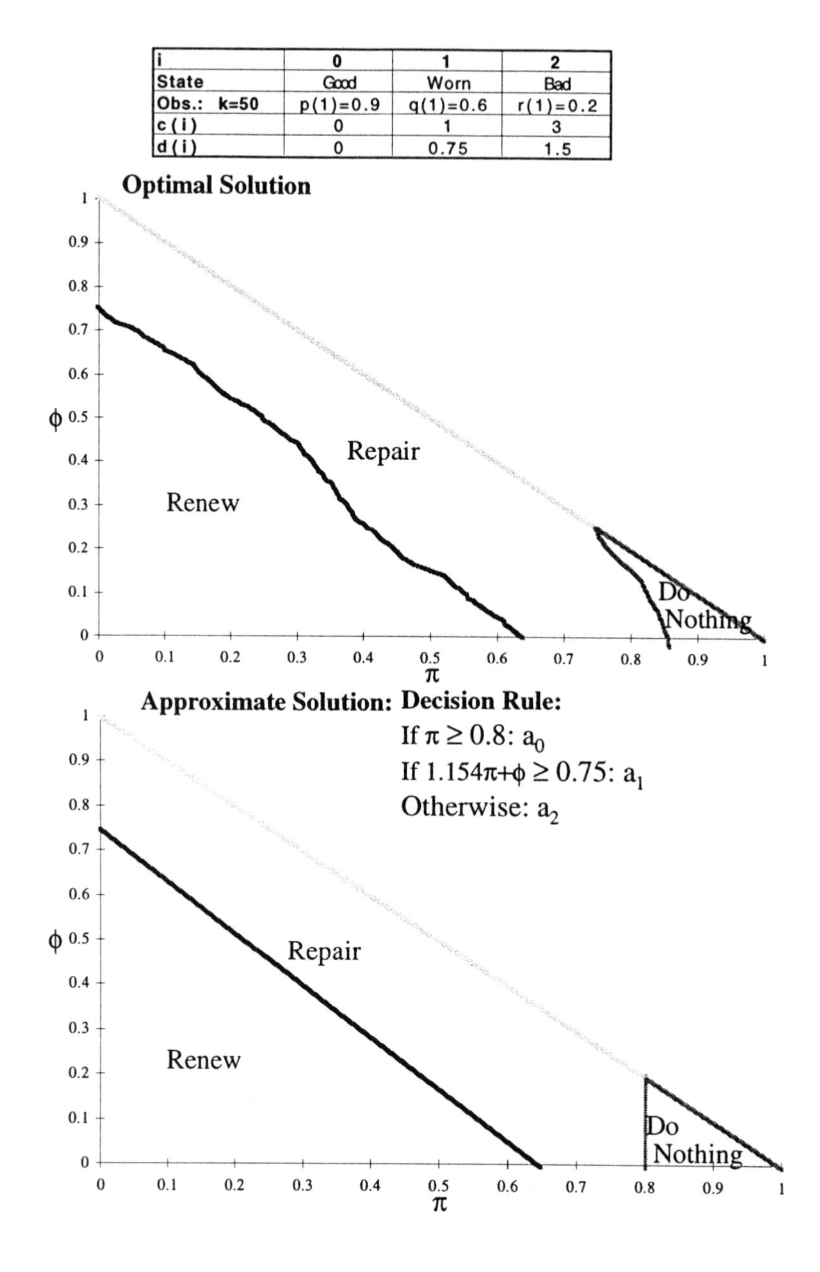

i	0	1	2
State	Good	Worn	Bad
Obs.: k=50	p(1)=0.9	q(1)=0.6	r(1)=0.2
c(i)	0	1	3
d(i)	0	0.75	1.5

FIGURE 7. Approximation for Decision Regions using a Linear Decision Rule

where we assume $0 < u \leq b \leq a < 1$ and a, b, and u are the same as in Section 4.1. The binomial distribution reasonably represents observations taken over a time interval to evaluate the condition of a system. For example, a manufacturer may

TABLE 3. Expected Cost per Unit Time for Optimal and Approximate Policies for $\pi = 1$ and $\phi = 0$ for Various Examples (one parameter is varied while the others are fixed to base case values).

Parameter Changed from Table 2	*Approximate Decision Rule*	*Optimal* $\frac{C^*_{50}(1,0)}{50}$	*Approximation* $\frac{C^a_{50}(1,0)}{50}$	*Percent Increase in Expected Cost*
$c_1 = 2.00$	If $\pi \geq 0.9$, a_0, If $\pi + \phi \geq 0.8$, a_1, Otherwise, a_2	0.4256	0.4292	0.85
$c_1 = 5.00$	If $\pi > 0.8$ and $\phi < 0.1$, a_0, If $\pi + \phi \geq 1.0$, a_1, Otherwise, a_2	0.4439	0.4449	0.23
$d_1 = 0.25$	If $\pi \geq 1.0$, a_0, If $\pi + \phi \geq 0.1$, a_1, Otherwise, a_2	0.1472	0.1483	0.75
$d_1 = 0.50$	If $\pi + \phi \geq 1.0$ and $\pi > 0.9$, a_0, If $1.25\pi + \phi \geq 0.5$, a_1, Otherwise, a_2	0.2744	0.2776	1.17
$d_2 = 1.75$	If $2\pi + \phi \geq 1.8$, a_0, If $\pi + \phi \geq 0.5$, a_1, Otherwise, a_2	0.3789	0.3821	0.84

measure a characteristic of a machine every 30 seconds but make a decision only every five minutes, based on the number of measurements in the group that satisfy a certain criterion. We can also let $n \to \infty$ to approximate a Normal distribution where $na \to \mu_1$, $nb \to \mu_2$, and $nc \to \mu_3$.

5.2.1. *Base Case.* Figure 9 shows the optimal decision regions for $k = 50$ and the parameters in Table 5. For the examples that follow, all parameters are set to the values in Table 5 except for the parameter being varied.

5.2.2. *Affect of Horizon and Sample Size on Decision Regions.* Similar to the example with Bernoulli observations, the optimal decision regions converge for $k \geq 5$. As we found in Section 4.1, the expected cost per unit time increases significantly as k increases from 1 to 5, converging numerically to a constant for values of k above 5.

The value of additional samples can be determined by comparing the expected cost per unit time as n increases. As shown in Fig. 10, for $n > 5$, the decision regions change only slightly with increasing n, the *do nothing* region gets a little larger, and the *repair* action is optimal for more values of ϕ. In fact, the decision regions become numerically stationary for $n \geq 15$.

As shown in Fig. 11, as the sample size increases from 1 to 20, the expected cost per unit time decreases by around ten percent. However, the decrease in the expected cost gets smaller as the sample size increases. In fact for $n \geq 20$, the

TABLE 4. Expected Cost per Unit Time for Two-State and Three-State Models for $\pi = 1$ and $\phi = 0$ as a function of γ, the two-state hazard where $\gamma = \frac{1}{\frac{1}{\beta_0} + \frac{1}{\beta_1}}$.

Two-State Hazard, γ	β_0	β_1	*Two-State Model*	*Three-State Model*	*Percent Increase in Expected Cost*
0.10	0.20	0.20	0.4588	0.3789	21.09
0.13	0.20	0.40	0.5508	0.4458	23.55
0.15	0.20	0.60	0.5935	0.4989	18.96
0.16	0.20	0.80	0.6132	0.5257	16.64
0.17	0.20	1.00	0.6221	0.5394	15.33
0.13	0.40	0.20	0.5508	0.4372	25.98
0.20	0.40	0.40	0.6666	0.5044	32.16
0.24	0.40	0.60	0.6973	0.5683	22.70
0.27	0.40	0.80	0.7150	0.6160	16.07
0.29	0.40	1.00	0.7265	0.6160	17.94
0.15	0.60	0.20	0.5935	0.4530	31.02
0.24	0.60	0.40	0.6973	0.5370	29.85
0.30	0.60	0.60	0.7380	0.5951	24.01
0.34	0.60	0.80	0.7380	0.6515	13.28
0.38	0.60	1.00	0.7380	0.6840	7.89
0.16	0.80	0.20	0.6132	0.4750	29.09
0.27	0.80	0.40	0.7150	0.5506	29.86
0.34	0.80	0.60	0.7380	0.6104	20.90
0.40	0.80	0.80	0.7440	0.6638	12.08
0.44	0.80	1.00	0.7440	0.7188	3.51
0.17	1.00	0.20	0.6221	0.4935	26.06
0.29	1.00	0.40	0.7265	0.5615	29.39
0.38	1.00	0.60	0.7380	0.6201	19.01
0.44	1.00	0.80	0.7440	0.6716	10.77
0.50	1.00	1.00	0.7500	0.7350	2.04

TABLE 5. Parameter Values for Binomial Monitoring Base Case.

i	*Deterioration Rate*	*State Operating Cost*	*Action Cost*	*Observation Distribution Parameter*
	β_i	c_i, *per unit time*	d_i, *fixed*	$(n = 20)$
0	0.2	0	0	$a = 0.9$
1	0.2	1	0.75	$b = 0.6$
2		3	1.50	$u = 0.2$

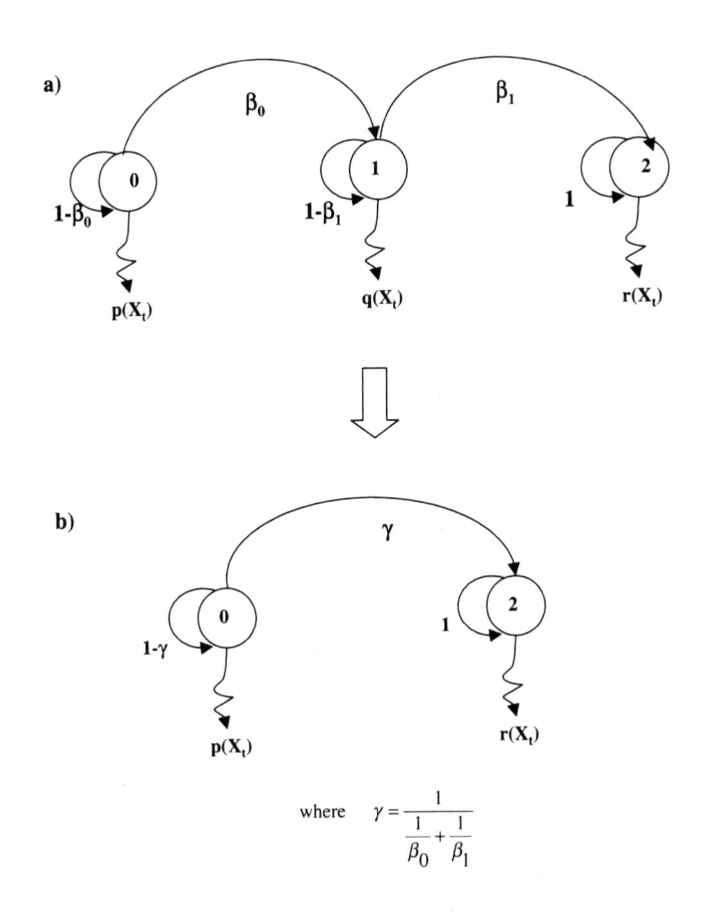

FIGURE 8. Three-State Model and the Equivalent Two-State Model.

decrease in expected cost is so small (i.e., less than 2 percent for this example) that it is almost insignificant. The larger the sample size, the less sensitive the expected cost per unit time is to increases in size. For this example, as shown by the example in Fig. 10, the shape of the decision regions converges for sample sizes between 10 and 15.

6. Summary

This paper focuses on the structure of optimal policies for maintenance or replacement decisions for deteriorating systems with probabilistic monitoring and "silent" failures. The goal is to characterize the structure of optimal policies to assist the decision maker in selecting maintenance and replacement actions over a finite horizon. The concept of "marginal monotonicity", which requires a componentwise partial ordering and monotonicity of the expected total cost function with respect to each component of the state space, allows characterization of the policy

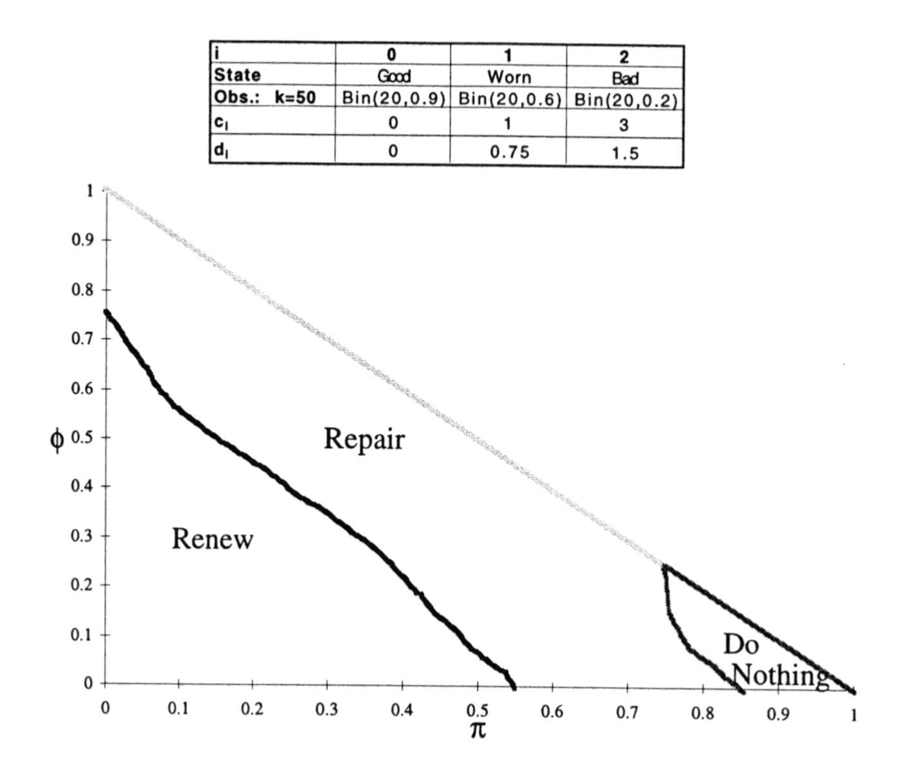

i	0	1	2
State	Good	Worn	Bad
Obs.: k=50	Bin(20,0.9)	Bin(20,0.6)	Bin(20,0.2)
c_i	0	1	3
d_i	0	0.75	1.5

FIGURE 9. Decision Regions for Binomial Monitoring Base Case.

that minimizes the total expected cost. This allows us to approximate the optimal policy by a collection of decision rules characterized by at most three functions. Extension of these results for an N-state process is presented in (16).

References

[1] S.C. Albright, *Structural results for partially observable Markov decision processes*, Operations Research **27** (1978), 1041–1053.

[2] C. Derman, *On sequential decisions and Markov chains*, Management Science **9** (1962), 16–24.

[3] L. Gong, W. Jwo, and K. Tang, *Using on-line sensors in statistical process control*, Management Science **43** (1997), no. 7, 1017–1028.

[4] W.J. Hopp and S.C. Wu, *Multiaction maintenance under Markovian deterioration and incomplete state information*, Naval Research Logistics **35** (1988), 447–462.

[5] ———, *Machine maintenance with multiple maintenance actions*, IIE Transactions **22** (1990), no. 3, 226–233.

[6] M. Klein, *Inspection-Maintenance-Replacement schedules under Markovian deterioration*, Management Science **9** (1962), 25–32.

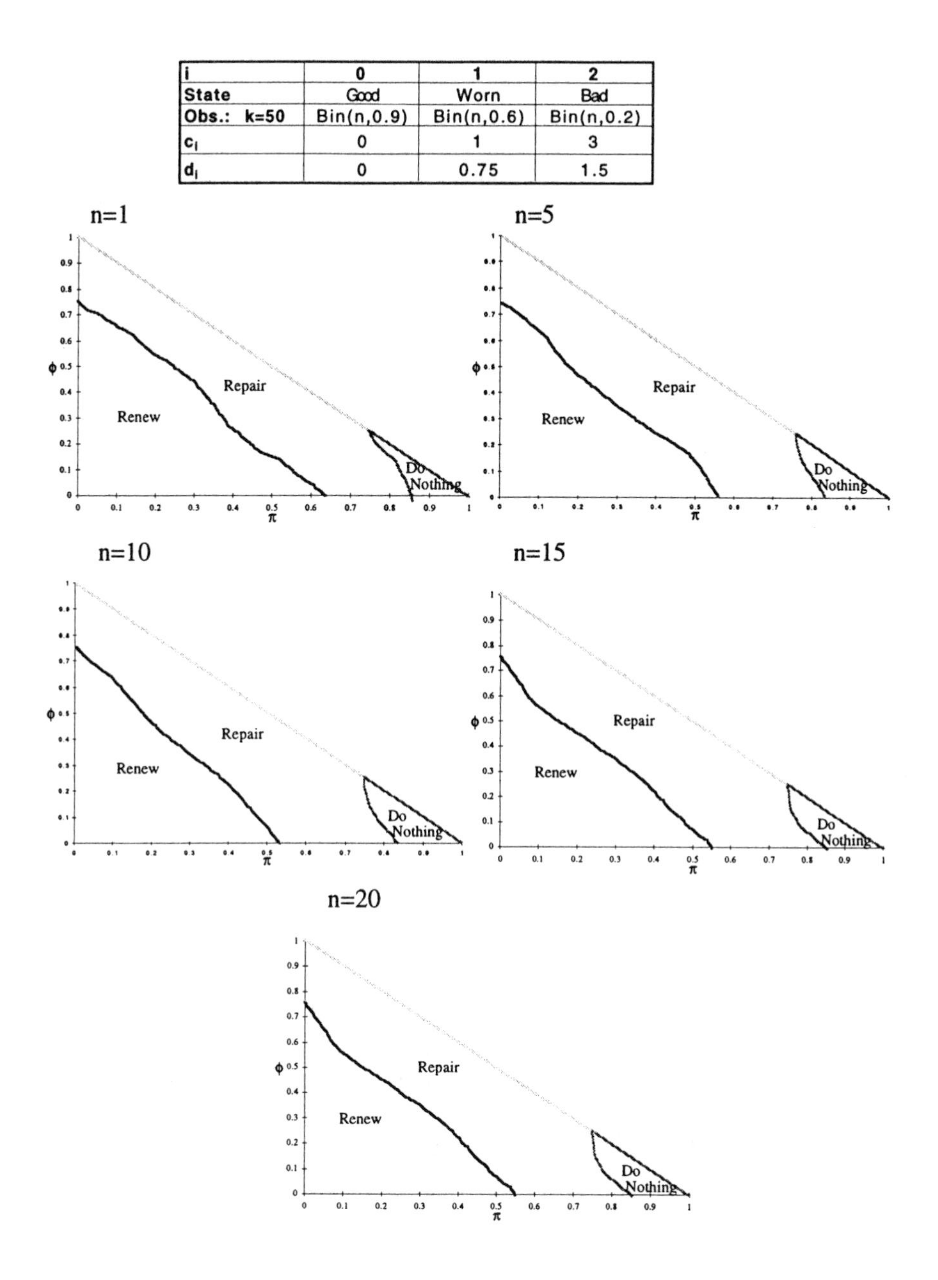

i	0	1	2
State	Good	Worn	Bad
Obs.: k=50	Bin(n,0.9)	Bin(n,0.6)	Bin(n,0.2)
c_i	0	1	3
d_i	0	0.75	1.5

FIGURE 10. Decision Regions as n increases.

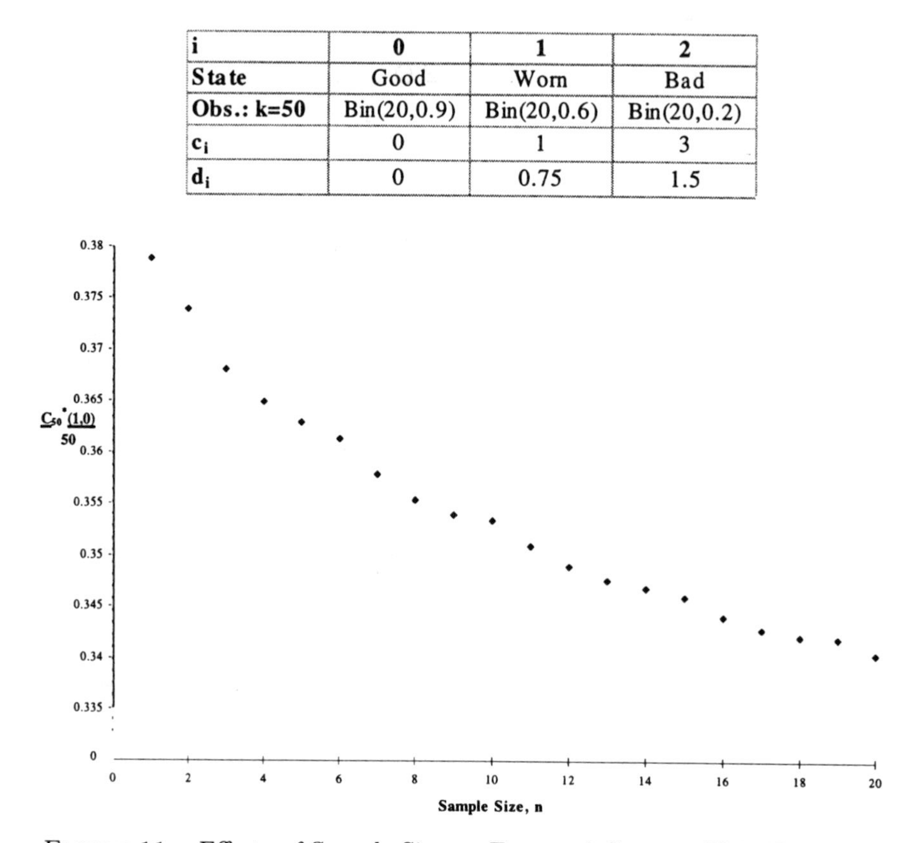

i	0	1	2
State	Good	Worn	Bad
Obs.: k=50	Bin(20,0.9)	Bin(20,0.6)	Bin(20,0.2)
c_i	0	1	3
d_i	0	0.75	1.5

FIGURE 11. Effects of Sample Size on Expected Cost per Unit Time.

[7] W.S. Lovejoy, *Some monotonicity results for partially observed Markov decision processes*, Operations Research **35** (1987), no. 5, 736–743.

[8] ———, *Computationally feasible bounds for partially observed Markov decision processes*, Operations Research **39** (1991), no. 1, 162–175.

[9] H. Luss, *Inspection policies for a system which is inoperable during inspection periods*, IIE Transactions **9** (1977), no. 2, 189–194.

[10] J.J. McCall, *Maintenance policies for stochastically failing equipment" a survey*, Management Science **11** (1965), no. 5, 493–524.

[11] G.E. Monahan, *A survey of partially observable Markov decision processes: theory, models, and algorithms*, Management Science **28** (1982), no. 1, 1–16.

[12] M. Ohnishi, T. Morioka, and T. Ibaraki, *Optimal minimal-repair and replacement problem of discrete-time markovian deterioration system under incomplete state information*, Computers and Industrial Engineering **27** (1994), no. 1-4, 409–412.

[13] W.P. Pierskalla and J.A. Voeller, *A survey of maintenance models: Control and surveillance of deteriorating systems*, Naval Research Logistics Quarterly

23 (1976), no. 3, 353–388.

[14] D. Rosenfield, *Markovian deterioration with uncertain information*, Operations Research **24** (1976), no. 1, 141=155.

[15] A.N. Shiryaev, *An optimum methods in quickest detection problems*, Theory of Applied Probability **8** (1968), 22–46.

[16] J.E. Simmons, *Maintenance and replacement policies for a multi-state deteriorating process with probabilistic monitoring*, Ph.D. thesis, Industrial and Operations Engineering Department, University of Michigan, Ann Arbor, 1998.

[17] R.D. Smallwood and E.J. Sondik, *The optimal control of partially observed Markov processes over a finite horizon*, Operations Research **21** (1973), 1071–1088.

[18] H.M. Taylor, *Markovian sequential replacement processes*, The Annals of Mathematical Statistics (1965), 1677–1683.

[19] D. Teneketzis and P. Varaiya, *The decentralized quickest detection problem*, IEEE Transactions on Automatic Control **AC-29** (1984), no. 7, 641–644.

[20] C. Valdez-Flores and R.M. Feldman, *A survey of preventative maintenance models for stochastically deteriorating single-unit systems*, Naval Research Logistics **36** (1989), 419–446.

[21] W. Wang, *Detection of process change with non-geometric failure time distribution*, Ph.D. thesis, Industrial and Operations Engineering Department, University of Michigan, Ann Arbor, 1995.

[22] C.C. White, III, *A Markov quality control process subject to partial observation*, Management Science **23** (1977), no. 8, 843–851.

[23] ———, *Optimal inspection and repair of a production process subject to deterioration*, Journal of Operational Research Society **29** (1978), no. 3, 235–243.

[24] ———, *Bounds on optimal cost for a replacement problem with partial observations*, Naval Research Logistics Quarterly **6** (1979), 415–422.

[25] ———, *Monotone control laws for noisy, countable state Markov chains*, European Journal of Operational Research **5** (1980), 124–132.

[26] C.C. White, III and D.J. White, *Markov Decision Processes*, European Journal of Operational Research **39** (1989), 1–16.

University of Michigan, Industrial and Operations Engineering, 1205 Beal Avenue, Ann Arbor, Michigan 48109-2117

Current address: University of Michigan Business School, 701 Tappan Street, Ann Arbor, Michigan 48109-1234

University of Michigan, Industrial and Operations Engineering, 1205 Beal Avenue, Ann Arbor, Michigan 48109-2117

E-mail address: `jsimmons@umich.edu`

Contemporary Mathematics
Volume **275**, 2001

The Application of Numerical Grid Generation to Problems in Computational Fluid Dynamics

Bonita V. Saunders

Abstract. Numerical grid generation, the computation of boundary fitted curvilinear coordinate systems to aid in the numerical solution of partial differential equations, is described. Grid generation plays a crucial role in resolving the problem of handling arbitrarily shaped boundaries when solving physical problems over a field. The driving impetus for the development of grid generation techniques was to solve problems in computational fluid dynamics, but grid generation is applicable to any area where partial differential equations are computed over a field. The use and benefits of grid generation are explained. Common types of grid generation systems are presented and finally, the author's work in generating grids suitable for solving physical problems that arise in solidification theory is described.

Introduction

In the field of computational fluid dynamics, the complexity of the physical problem often means that any realistic mathematical model must be solved numerically on a computer. One of the most time consuming tasks can be determining and constructing the coordinate system for the computations. A common technique, called numerical grid generation, is to develop a general curvilinear coordinate system that maps the oddly shaped physical domain back to a simpler computational domain such as a square or rectangle. This technique is also known as boundary-fitted or boundary conforming grid generation because the boundary of the mesh, or grid, generated by the coordinate system coincides with the boundary of the physical domain as shown in Figure 1.

Numerical grid generation has been an active area of research for many years, but the bulk of the research has been conducted during the last thirty years [1-21]. A great deal of progress has been made, but there is still more work to be done. Although grids can now be made for most boundary configurations, in many cases, especially in three dimensions, the process is neither easy nor automatic. Slight changes in a configuration can cause a lot of additional work. Many engineers complain that in the development of flow solver codes, the generation of the mesh continues to be the most time consuming part of the calculation [**3**]. Furthermore, the interface that connects the grid generator to the flow solver code is often hard

1991 *Mathematics Subject Classification.* Primary 65L50; Secondary 35R35.

Key words and phrases. Numerical grid generation, adaptive grid generation, free/moving boundary problems, solidification modeling .

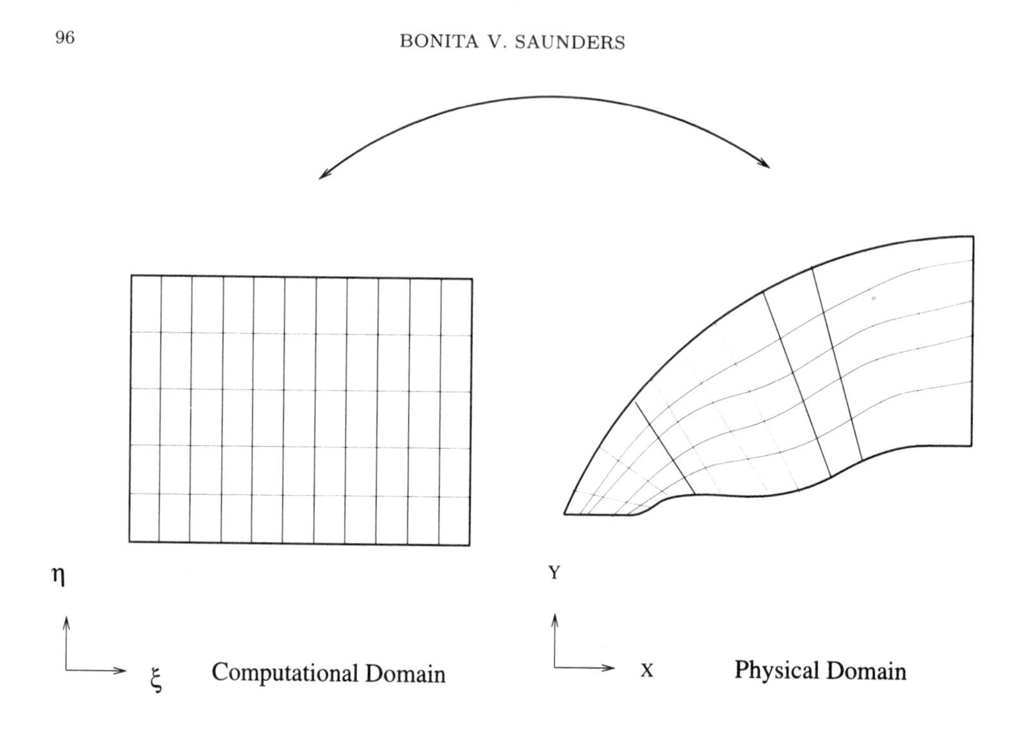

FIGURE 1. Boundary-fitted grid generation.

to use and too restrictive [**4**]. The need for adaptive codes continues to drive a lot of research. In the area of free and moving boundary problems, researchers continue to look for ways of developing grids that easily adapt to rapid and severe changes in a boundary without degrading the accuracy of the computations of the flow solver. These are just a few of the problems facing researchers in grid generation.

This paper looks at the motivation behind the development of grid generation systems, presents a brief introduction to the field, examines the use of grids, and discusses common types of grid generation systems. It also looks at the author's current interests in the field and, in particular, examines the challenge of creating grids that model a special type of free/moving boundary problem that arises in the field of solidification theory.

The Use of Numerical Grid Generation

The mathematical modeling of many physical processes, such as airflow around a wing or fuselage, or fluid flow around a ship, involves the solving of partial differential equations over an oddly shaped field. This makes it difficult to apply numerical solution techniques without introducing undesirable errors into the calculations. Numerical grid generation permits the user to transform the oddly shaped domain to a simpler domain on which it is easier to compute. Numerical grid generation is actually the creation of a curvilinear coordinate system that connects a simpler computational domain, such as a square or rectangle, to the more complicated physical domain as was seen in Figure 1. It is commonly called boundary-fitted or boundary conforming grid generation because the boundary of the mesh, or grid, the system generates on the physical domain matches the physical boundary. Partial differential equations and boundary conditions originally defined on the physical domain are transformed to equations on the simpler domain. The new equations tend to look more complicated, but the natural structure of the computational domain simplifies the coding of finite difference or finite element equations. Boundary conditions are now simple to apply because in the computational domain the boundary points lie on the boundary of a square or rectangle.

The default method of simply placing a rectangular mesh on the physical domain complicates the computation of boundary conditions because the grid may overlap the boundary in some areas as shown in Figure 2. Since no mesh point

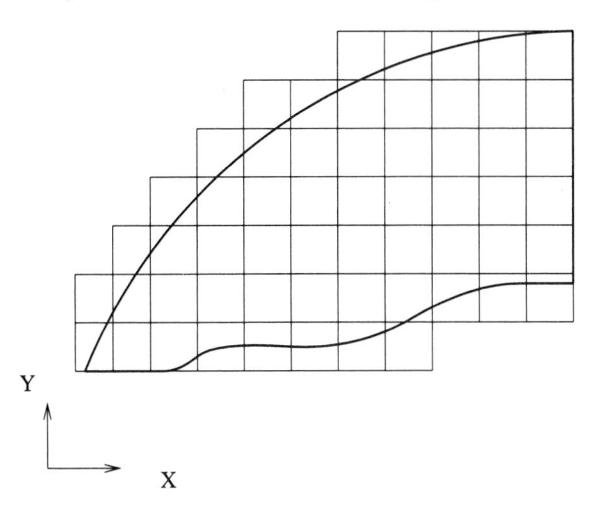

FIGURE 2. Cartesian mesh over physical domain.

lies directly on the boundary in such an area, interpolation is typically used to compute the boundary conditions there. This introduces errors into the numerical computations.

Another common technique for dealing with physical boundaries of arbitrary shape is to use triangulation methods. Many find these methods more adept at handling boundaries that have unusual shapes, but such methods generally require more memory because the user must store connectivity information, that is, information about how the mesh points relate to each other and which are neighbors of a given point. Also, even though triangulation methods may appear to fit a boundary well, the user may have trouble making the triangles small enough to get the resolution needed for accurate calculations near the boundaries. This is particularly true for viscous flow simulations needed to study flow around an airplane wing. A fine concentration of grid points primarily in the direction normal to the wing is needed to obtain accurate calculations of the boundary layer near the wing surface [**5**].

For all types of grid generation techniques a considerable amount of research continues to be devoted to the development of adaptive techniques in which grid points are either redistributed or added and deleted in response to what happens as the solution evolves [**6**]. Some methods capture gradient information and redistribute points in areas where the gradient is large. In the case of free and moving boundary problems, grid points are redistributed or added and deleted to follow the motion of the changing boundary.

Types of Grid Generation Systems

The most common types of numerical grid generation systems are partial differential equation generated systems, algebraically generated systems, and systems generated by variational methods. Partial differential equation systems include conformal, elliptic, parabolic and hyperbolic. Such systems tend to produce smooth grids, but they introduce the complication of solving additional partial differential equations besides those governing the physical problem. The use of conformal mapping techniques is probably the oldest method for constructing coordinate systems. In conformal systems the curvilinear coordinates can be generated by solving Laplace's equation with the Cauchy Riemann equations as the boundary conditions. If ξ,η are the curvilinear coordinates and x,y, the cartesian coordinates, then we have

$$\nabla^2\xi = \frac{\partial^2\xi}{\partial x^2} + \frac{\partial^2\xi}{\partial y^2} = 0 \tag{0.1}$$

$$\nabla^2\eta = \frac{\partial^2\eta}{\partial x^2} + \frac{\partial^2\eta}{\partial y^2} = 0$$

with boundary conditions

$$\frac{\partial\xi}{\partial x} = \frac{\partial\eta}{\partial y} \tag{0.2}$$

$$\frac{\partial\xi}{\partial y} = -\frac{\partial\eta}{\partial x}. \tag{0.3}$$

The equations and boundary conditions are transformed to the computational domain and solved. The orthogonality of conformal systems helps minimize truncation error in finite difference calculations, but conformal systems permit little control over grid spacing if grid points need to be concentrated in certain areas to obtain accuracy. A conformal system is actually a special type of elliptic grid generation system.

Elliptic systems go back at least thirty years. Winslow [**7**] was one of the early users, but elliptic systems became popular in the 1970s and 1980s when they were reintroduced and improved by Joe Thompson and Wayne Mastin *et al* of Mississippi State University [**2, 8**]. Elliptic systems are generated from either the Laplace equations or the Poisson equations obtained by adding functions P and Q that control the spacing of the coordinate lines:

$$\nabla^2\xi = \frac{\partial^2\xi}{\partial x^2} + \frac{\partial^2\xi}{\partial y^2} = P \tag{0.4}$$

$$\nabla^2\eta = \frac{\partial^2\eta}{\partial x^2} + \frac{\partial^2\eta}{\partial y^2} = Q.$$

Elliptic systems produce very smooth grids, but choosing the appropriate grid control functions, determining the best way to match a complicated physical boundary with the computational boundary, and solving the elliptic system can be a time consuming and tricky process.

S. Nakamura used parabolic partial differential equations to generate coordinate systems [**9**] and Steger and colleagues [**10**] worked on hyperbolic grid generation systems. Such systems are generally faster than elliptic, but they are only applicable to physically unbounded regions [**11**].

Inherent in any grid generation system is an invertible mapping of the curvilinear coordinates onto the cartesian coordinates. In systems generated by partial differential equations the mapping is not directly available, but in algebraic generation systems the mapping is given explicitly. Hence no partial differential equations need to be solved to generate the curvilinear coordinate system. One of the earliest and simplest algebraic generation techniques is transfinite, or blending function, interpolation. In this type of system the grid generation mapping is actually an interpolation function that is described in terms of functions specified on the boundary. More complex transfinite systems can be created by also specifying functions on selected interior curves. William Gordon *et al* did a substantial amount of work on these systems at General Motors during the 60s, 70s, and 80s [**12, 13**]. Others who have worked with algebraic systems are R.E. Smith [**14**], P.R. Eiseman [**15**], and B.V. Saunders [**16**]. Generally, algebraic systems are faster grid generators, but the grids tend not to be as smooth. Singularities on the boundary may propagate into the interior of the grid. Some of the problems with algebraic systems can be lessened or eliminated by applying smoothing techniques. In a later section of this paper an algebraic system designed by the author which uses variational techniques for smoothing is described.

Brackbill and Saltzman [**17**] and Steinberg and Roache [**18**] popularized variational grid generation methods. Grids are generated by solving the Euler-Lagrange equations derived by minimizing three integrals that control grid smoothness, orthogonality, and the area of grid cells. J. Castillo [**19**] developed a discrete variational system that uses sums rather than integrals.

Many grid generation systems consist of a combination of several systems. For example, an algebraic system may be used to obtain an initial grid and an elliptic or variational technique used to smooth it. One of the most effective combination grid systems is the multi-block or block-structured system which was introduced in the 1980s. Multi-block systems are formed by dividing the physical domain into several simpler sections or blocks. A grid generation system is then designed for each block. Using multi-block systems, grids have been created for very complicated three dimensional configurations [**1**].

Applications of Numerical Grid Generation

Much of the initial research in numerical grid generation was motivated by a desire to solve problems in computational fluid dynamics. Early systems were used to model aerodynamic and hydrodynamic phenomena such as airflow around an airplane wing or fuselage, airflow around a moving automobile, or fluid flow around a ship or submarine. Over the years the use of grid generation has expanded into nontraditional areas such as the modeling of flow through porous media and the modeling of the solidification of materials, a field which involves the study of fluid flow as well as both heat and mass transfer. Grid generation is also applicable to problems in electromagnetism, structures, and any other area involving physical phenomena that can be modeled by the solution of differential equations over a field.

Figure 3 shows a two-dimensional grid around an airfoil, that is, a cross section of a wing. Such a grid would be used to solve the Navier-Stokes equations of motion around the airfoil. The solution data helps engineers judge the effectiveness of the airfoil shape. The boundary data was provided by R.E. Smith of NASA Langley Research Center. The grid was generated using an algebraic system developed by the author. The system, which uses a mapping composed of tensor product B-splines, is described in detail in [**16**], but a modified version designed to interpolate an interior interface is described in the next section. Note that the grid

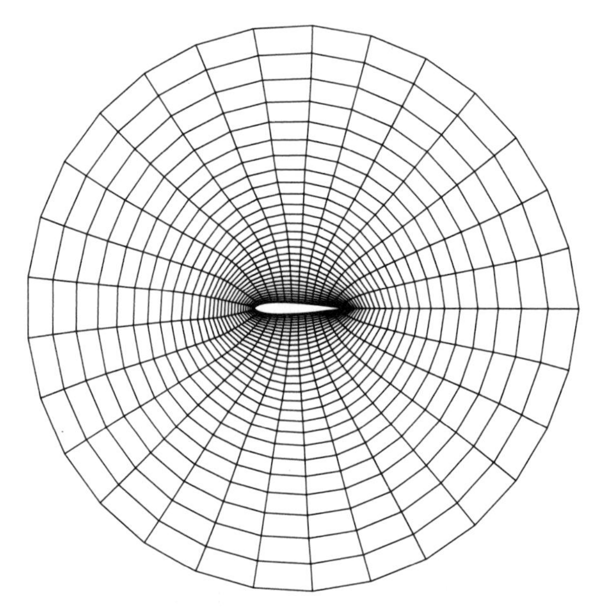

FIGURE 3. Airfoil grid.

is concentrated near the boundary of the airfoil because more points are needed to get an accurate picture of what the flow lines look like in that area. Farther away the air flow is less affected by the body. Hence, the flow tends to be very smooth and uninteresting and fewer grid points are needed for accurate calculations. Also notice that over much of the mesh, the grid lines are close to orthogonal. A large degree of nonorthogonality severely increases the truncation error in finite difference calculations.

Another area where challenging grid generation problems arise is in the study of solidification theory. Understanding the microstructures that develop during the process of solidification can help metallurgists improve the quality of metal products manufactured by casting or welding, or aid solid-state physicists in producing pure semiconductor crystals needed for electronic devices. A common technique used to study solidification is Bridgman growth, a directional solidification method in which a small sample of the metal alloy is drawn through a constant temperature gradient at a uniform rate of speed, V, as shown in Figure 4.

Mullins and Sekerka discovered that there is a critical velocity at which the solid-liquid interface will become unstable [**22**]. As the growth speed is increased, the original flat or planar interface deforms to a sinusoidal shape, then to a bulb-like cellular shape, and then to a dendritic shape as shown in Figure 5. To fully

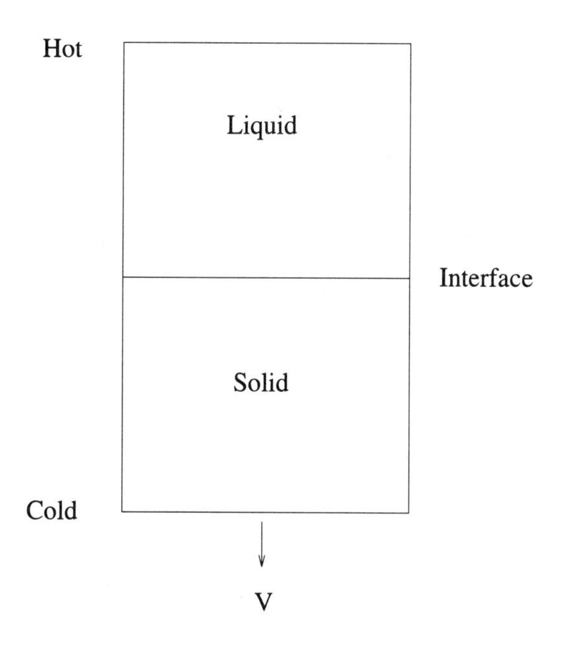

FIGURE 4. Bridgman growth technique.

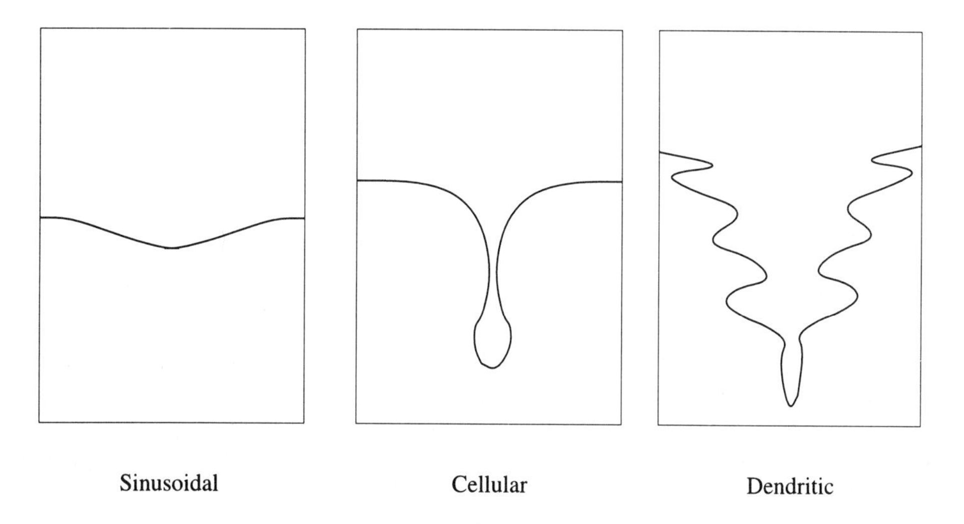

FIGURE 5. Instability of solid-liquid interface.

understand this process, researchers conduct experiments or model the process numerically. Ideally, to model the process, one should use an adaptive grid with interior grid lines that conform to the shape of the interface as it deforms. However, even tracking the deformation to the cellular shape can be quite difficult because the cells can become very deep and narrow with re-entrant bulb-like shapes as the control parameters, either growth velocity or temperature gradient, are modified. The grid must adapt to severe deformations while maintaining as much smoothness and orthogonality as possible. Dendritic shapes are even harder to track. Consequently, in the study of complex dendritic shapes, a considerable amount of research has been devoted to phase field models where the interface is not tracked explicitly [**23, 24, 25**]. Yet even in such models, grid concentration in the general area of the interface is beneficial.

The next section describes a boundary-fitted grid generation system that the author is developing to be used to model the Bridgman growth of a binary alloy. It is designed to track the interface of the solidifying alloy as it deforms from a planar shape to a deeply grooved cell. Brown and colleagues at MIT have done quite a bit of research in this area [**26, 27, 28**], but the grids they developed for deep cells required that the interface be divided into sections [**27**], or that two procedures be used, one for each coordinate direction [**28**]. The system described in this paper requires no division of the interface or domain. Furthermore, it is an algebraic system which means no partial differential equations must be solved to obtain the grid. For completeness, the mapping and some examples are described in the next section, but a more detailed discussion is presented in [**20**].

A Grid Generation Mapping for Solidification Modeling

To facilitate the modeling of Bridgman growth, the boundary fitted grid generation mapping should fit the interface curve on the interior of the physical domain in addition to fitting the rectangular outer boundary. Furthermore, the system should be adaptive since the grid lines must change to follow the deforming interface while maintaining as much smoothness and orthogonality as possible. Therefore, we design a mapping, $\mathbf{T}$, that maps the unit square, I_2, onto the physical domain and is constructed so that the interface is the coordinate curve $\eta = 1/2$ as shown in Figure 6. The mapping has the form

$$\mathbf{T}(\xi,\eta) = \begin{pmatrix} x(\xi,\eta) \\ y(\xi,\eta) \end{pmatrix} = \begin{pmatrix} \sum_{i=1}^{m}\sum_{j=1}^{n} \alpha_{ij}B_{ij}(\xi,\eta) \\ \sum_{i=1}^{m}\sum_{j=1}^{n} \beta_{ij}B_{ij}(\xi,\eta) \end{pmatrix}, \tag{0.5}$$

where $0 \le \xi, \eta \le 1$ and $B_{ij}(\xi,\eta) = B_i(\xi)B_j(\eta)$ where B_i and B_j are elements of cubic B-spline sequences associated with finite nondecreasing knot sequences, $\{s_i\}_1^{m+4}$ and $\{t_j\}_1^{n+4}$, respectively.

The spline coefficients can be divided into three groups. The boundary coefficients, α_{1j}, β_{1j} and α_{mj}, β_{mj} for $j = 1, \ldots, n$ and α_{i1}, β_{i1} and α_{in}, β_{in} for $i = 1, \ldots, m$, are the coefficients of the B_{ij} that are nonzero on the boundary of I_2. The inteface coefficients are the coefficients of the B_{ij} that might be nonzero when $\eta = 1/2$. They are α_{ik}, β_{ik} where $1 \le i \le m$, $l-3 \le k \le l$ and l is such that $1/2 \in [t_l, t_{l+1}]$. The remaining coefficients are called the interior coefficients.

Initially, the coefficients are chosen to approximate a transfinite blending function interpolant that matches the outer boundary and interface of the physical domain. To decrease skewness and increase the smoothness of the grid lines the initial coefficients are modified to minimize the functional

$$F = \int_{I_2} \left[w_1 \left\{ \left(\frac{\partial J}{\partial \xi}\right)^2 + \left(\frac{\partial J}{\partial \eta}\right)^2 \right\} + w_2 \left\{ \frac{\partial \mathbf{T}}{\partial \xi} \cdot \frac{\partial \mathbf{T}}{\partial \eta} \right\}^2 \right] dA \tag{0.6}$$

where $\mathbf{T}$ denotes the grid generation mapping, J is the Jacobian of the mapping, and w_1 and w_2 are weight constants. When w_1 is large, the variation of the Jacobian values at nearby points will be small, thereby increasing the smoothness of the grid. When w_2 is large, the dot product term will be small, causing the grid lines to approach orthogonality. To avoid solving the Euler-Lagrange equations for the

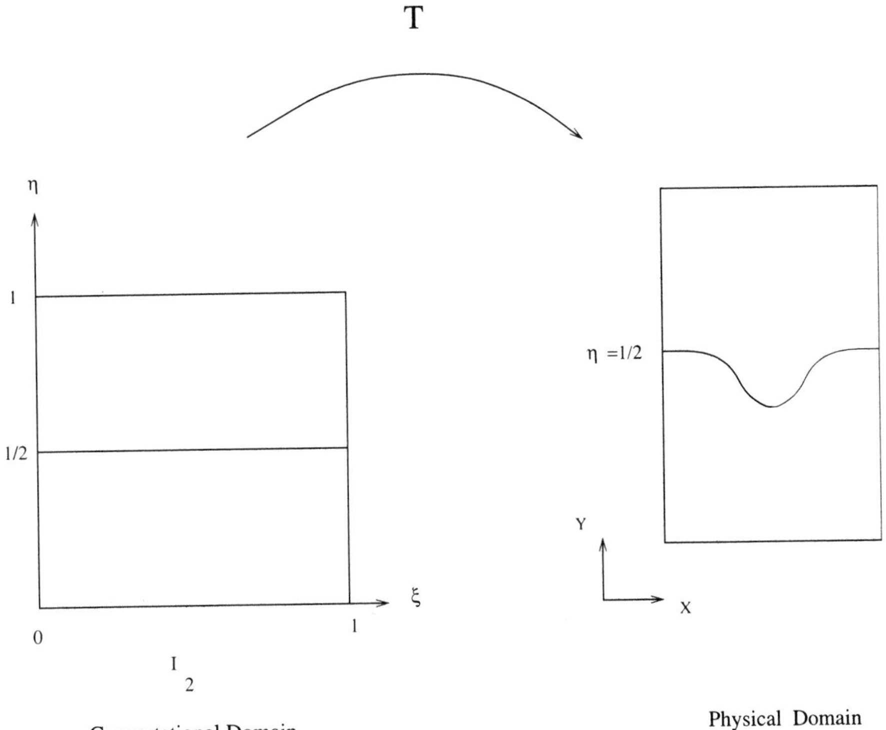

FIGURE 6. Grid generation mapping.

variational problem, we approximate this functional in the computer code by using the sum

$$
\begin{aligned}
G &= \sum_{i,j} w_1 \left[\left(\frac{J_{i+1,j} - J_{ij}}{\triangle\xi} \right)^2 + \left(\frac{J_{i,j+1} - J_{ij}}{\triangle\eta} \right)^2 \right] \triangle\xi\triangle\eta \\
&\quad + \sum_{i,j} w_2 Dot_{ij}^2 \triangle\xi\triangle\eta
\end{aligned}
\tag{0.7}
$$

where J_{ij} is the Jacobian value and Dot_{ij} is the dot product of $\partial\mathbf{T}/\partial\xi$ and $\partial\mathbf{T}/\partial\eta$ at mesh point (ξ_i, η_j) on the unit square. Surprisingly, G is a fourth degree polynomial in each spline coefficient so the minimum is found by using a cyclic coordinate descent technique which sequentially finds the minimum with respect to each coefficient. The minimization code takes advantage of the small support of B-splines when evaluating the sums that comprise G and is highly vectorizable.

Figures 7 and 8 show the system's ability to generate grids that conform to extremely deformed cellular shapes that are typical of experimental and numerical results seen to date. The grids in Figure 7 are for a sinusoidal interface. The first grid was obtained by choosing spline coefficients to approximate a transfinite interpolation mapping. The second grid shows the improved grid obtained after the coefficients are modified to minimize the smoothing functional G. The system untangles and smoothly distributes the grid lines underneath the interface. The grids in Figure 8 show the grid system's ability to maintain a significant amount of smoothness and orthogonality, while adapting to a very deep interface.

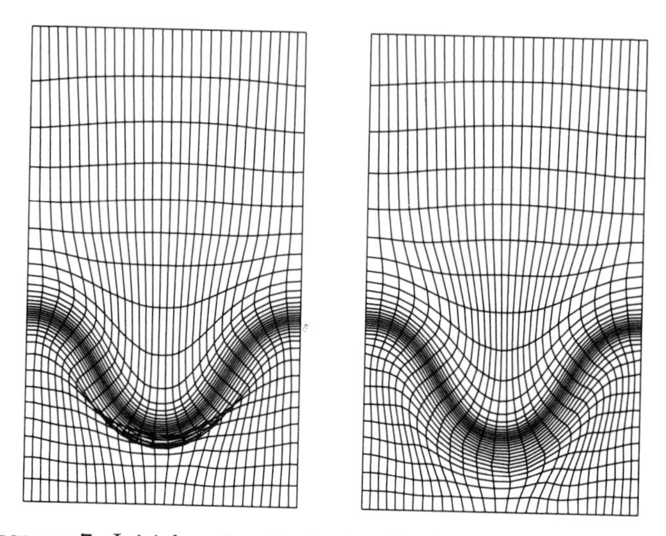

FIGURE 7. Initial and optimized grids for mildly deformed sinusoidal shaped interface.

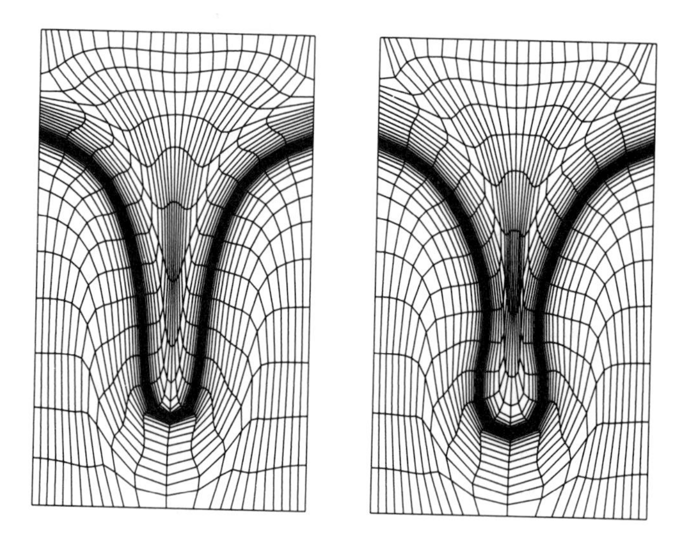

FIGURE 8. Optimized grids for deep and re-entrant cellular interfaces.

The final phase of this work is the coupling of the grid generation software with a solver for the system of partial differential equations that model the distribution of solute in the Bridgman growth process. However, it is clear that testing and improvement of the grid generation software will continue even as the author begins to focus more on the development of the solver.

Conclusions and Comments

We have presented an overview of numerical grid generation, briefly described its use, and looked at some areas of application. Numerical grid generation can be a very effective tool in removing the complication of shape from the modeling of physical phenomena over a field. A great deal of progress has been made over the last thirty years, but more work is needed in several areas such as grid adaption, the development of flexible codes, and the development of better interfaces for flow solvers and geometric modelers. Also, the teaching of grid generation techniques needs to be more widespread so that the field is more accessible to potential users rather than grid generation specialists. Furthermore, even though the original motivation for grid generation came from the field of computational fluid dynamics, grid generation is now used by researchers in many fields. One such area is illustrated by the author's work in the field of solidification theory. A concerted effort should be made to encourage the attendance of researchers in other fields at grid generation conferences. At the same time, grid generation specialists should seek out other areas of application and present their work at conferences in other fields.

References

[1] Thompson, J. F. - A Reflection on Grid Generation in the 90s: Trends, Needs, and Influences, Numerical Grid Generation in Computational Field Simulations, ed. by B.K. Soni *et al*, pp. 1029-1110, Mississippi State University, Mississippi, 1996.

[2] Thompson, J. F., Warsi, Z. U. A., and Mastin, C. W. - Boundary-fitted Coordinate Systems for Numerical Solution of Partial Differential Equations: A Review, Journal of Computational Physics, Vol. 47, pp. 1-108, 1982.

[3] Connell, S.D., Sober, J.S. and Lamson, S.H. - Grid Generation and Surface Modeling for CFD, Proceedings of the Surface Modeling, Grid Generation, and Related Issues in Computational Fluid Dynamics Workshop, NASA Conference Publication 3291, p. 29, NASA Lewis Research Center, Cleveland, Ohio, 1995.

[4] Cosner, R.R. - Future Requirements in Surface Modeling and Grid Generation, Proceedings of the Surface Modeling, Grid Generation, and Related Issues in Computational Fluid Dynamics Workshop, NASA Conference Publication 3291, p. 3, NASA Lewis Research Center, Cleveland, Ohio, 1995.

[5] Kallinderis, Y. - Discretization of Complex 3-D Flow Domains with Adaptive Hybrid Grids, Numerical Grid Generation in Computational Field Simulations, ed. by B.K. Soni *et al*, pp. 505-515, Mississippi State University, Mississippi, 1996.

[6] Hawken, D.F., Gottlieb, J.J., and Hansen, J.S. - Review of Some Adaptive Node-Movement Techniques in Finite-Element and Finite-Difference Solutions of Partial Differential Equations, Journal of Computational Physics, Vol. 95, pp. 254-302, 1991.

[7] Winslow, A.M. - Numerical Solution of the Quasilinear Poisson Equation in a Nonuniform Triangle Mesh, Journal of Computational Physics, Vol. 2, pp. 149-172, 1967.

[8] Thompson, J.F., Thames, F.C., and Mastin, C.W. - Automatic Numerical Generation of Body-Fitted Curvilinear Coordinate System for Field Containing Any Number of Arbitrary Two-Dimensional Bodies, Journal of Computational Physics, Vol. 15, pp. 299-319, 1974.

[9] Nakamura, S. - Marching Grid Generation Using Parabolic Differential Equations, Numerical Grid Generation, ed. by J.F. Thompson, North-Holland, pp. 775-807, 1982.

[10] Steger, J.L. and Sorenson, R.L. - Use of Hyperbolic Partial Differential Equations to Generate Body Fitted Coordinates, Numerical Grid Generation Techniques, ed. by R.E. Smith, pp. 463-478, NASA-CP-2166, 1980.

[11] Thompson, J. F., Warsi, Z. U. A., and Mastin, C. W. - Numerical Grid Generation: Foundations and Applications, North-Holland, 1985.

[12] Gordon, W.J. - Spline-Blended Surface Interpolation Through Curve Networks, Journal of Mathematics and Mechanics, Vol. 18, No.10, pp. 931-952, 1969.

[13] Gordon, W.J. and Hall, C.A. - Construction of Curvilinear Coordinate Systems and Applications to Mesh Generation, International Journal for Numerical Methods in Engineering, Vol. 7, pp.461-477, 1973.

[14] Smith, R.E. - Algebraic Grid Generation, Numerical Grid Generation, ed. by J.F. Thompson, North-Holland, 1982.

[15] Eiseman, P.R. - A Multi-Surface Method of Coordinate Generation, Journal of Computational Physics, Vol. 33, pp. 118-150, 1979.

[16] Saunders, B.V. - Algebraic Grid Generation Using Tensor Product B-splines, NASA CR-177968, 1985.

[17] Brackbill, J.U. and Saltzman, J.S. - Adaptive Zoning in Singular Problems in Two Dimensions, Journal of Computational Physics, Vol. 46, pp. 342-368, 1982.

[18] Steinberg, S. and Roache, P.J. - Variational Grid Generation, Numerical Methods for Partial Differential Equations, Vol. 2, pp. 71-96, 1986.

[19] Castillo, J.E. - Discrete Variational Grid Generation, Mathematical Aspects of Numerical Grid Generation, ed. by J.E. Castillo, pp. 35-48, 1991.

[20] Saunders, B.V. - A Boundary Conforming Grid Generation System for Interface Tracking, Computers and Mathematics with Applications, Vol. 29, No. 10, pp. 1-17, 1995.

[21] Saunders, B.V. - A Boundary-Fitted Grid Generation System for Interface Tracking, Numerical Grid Generation in Computational Field Simulations, ed. by B.K. Soni it et al, pp. 599-608, Mississippi State University, Mississippi, 1996.

[22] Mullins, W.W. and Sekerka, R.F. - Stability of a Planar Interface During Solidification of a Dilute Binary Alloy, Journal of Applied Physics, Vol. 35, No. 2, pp. 444-451, 1964.

[23] Wheeler, A.A., Boettinger, W.J. and McFadden, G.B. - Phase-Field Model for Isothermal Phase Transitions in Binary Alloys, Physical Review A, Vol. 45, No. 10, pp. 7424-7439, 1992.

[24] Wheeler, A.A., Murray, B.T. and Schaefer, R.J. - Computation of Dendrites Using a Phase Field Model, Physica D, Vol. 66, pp. 243-262, 1993.

[25] Wang, S.-L., Sekerka, R.F., Wheeler, A.A., Murray, B.T., Coriell, S.R., Braun, R.J. and McFadden, G.B. - Thermodynamically-Consistent Phase-Field Models for Solidification, Physica D, Vol. 69, pp. 189-200, 1993.

[26] Ettouney, H.M. and Brown, R.A. - Finite-Element Methods for Steady Solidification Problems, Journal of Computational Physics, Vol. 49, pp. 118-150, 1983.

[27] Ungar, L.H., Bennett, M.J. and Brown, R.A. - Cellular Interface Morphologies in Directional Solidification. IV. The Formation of Deep Cells, Phys. Rev. B, Vol. 31, No. 9, pp. 5931-5940, 1985.

[28] Tsiveriotis, K. and Brown, R.A. - Boundary-Conforming Mapping Applied to Computations of Highly Deformed Solidification Interfaces, Int. J. Numer. Methods Fluids, Vol. 14, pp. 981-1003, 1992.

National Institute of Standards and Technology, 100 Bureau Drive, Stop 8910, Gaithersburg, MD 20899-8910

E-mail address: `bonita.saunders@nist.gov`

Contemporary Mathematics
Volume **275**, 2001

The Elementary Residual Method

Desmond Stephens and Gary Howell

ABSTRACT. This paper introduces the ELeMentary RESidual method (ELMRES) for solving sparse systems of linear equations. ELMRES is an oblique projection method which generates the basis vectors for the Krylov subspace using the Hessenberg algorithm. Orthogonal projection methods such as Generalized Minimal Residual (GMRES) and its variants use modified Gram-Schmidt or Householder transformations to construct an orthogonal basis for the Krylov subspace. ELMRES constructs a basis for the same subspace in half the number of operations as GMRES. ELMRES minimizes the elementary residual, $\|L_{m+1}^{-1}(b - Ax_m)\|_2$.

1. Introduction

Krylov Subspace methods are used to solve large sparse linear systems,

$$Ax = b \tag{1.1}$$

where A is an $n \times n$ matrix. The ELeMentary RESidual method (ELMRES) is motivated by Generalized Minimal Residual (GMRES) [**D**]. Both of these methods are Krylov subspace methods. GMRES is an orthogonal projection method which forms an orthonormal basis for the Krylov subspace,

$$K_m = span\{x, Ax, A^2x, \dots, A^{m-1}x\} \tag{1.2}$$

in exact arithmetic. ELMRES is an oblique projection method which forms a basis for K_m using elementary similarity transformations. GMRES, which forms the basis using modified Gram-Schmidt, takes twice as many operations as ELMRES. In some cases the basis vectors generated by GMRES lose orthogonality requiring that they be reorthogonalized. If orthogonality is paramount, Householder transformations may be employed. A comparison of the operation counts of ELMRES and GMRES can be found in Table 1. In our experience convergence behavior of ELMRES and GMRES has been comparable. ELMRES has proven to be just as robust as GMRES. Restarts and preconditioning influence both algorithms in similar ways.

Section 2 describes GMRES for comparison to ELMRES. Section 3 describes Hessenberg's algorithm for constructing a basis for the Krylov subspace. We show that this algorithm is equivalent to reduction of A to Hessenberg form by elementary similarity transformations. Section 4 gives a derivation of the elementary residual

1991 *Mathematics Subject Classification.* Primary 65F10.
Key words and phrases. Krylov Subspaces, Oblique Projection Methods, GMRES .

TABLE 1. **Comparison of flops for GMRES with Modified Gram-Schmidt (MGS), reorthogonalization (MGSR), Householder (HO) and ELMRES using elementary transformations.**

GMRES (MGS)	GMRES (MGSR)	GMRES (HO)	ELMRES
$2m^2n$	$4m^2n$	$4m^2n - 4m^3/3$	$m^2n - m^3/3$

and Section 5 reviews our numerical results. In Section 6 we give conclusions and outline possible future work.

2. Overview of GMRES

We briefly review GMRES so that it can be compared to ELMRES. The procedure for GMRES is as follows [**D**], [**E**]. Let x_0 be an initial guess of a solution to (1.1) and r_0 be the residual such that $r_0 = b - Ax_0$. The orthonormal basis vectors for (1.2),

$$V_m = [v_1, v_2, \dots, v_m]$$

are constructed by first letting $v_1 = \frac{r_0}{\|r_0\|_2}$. The size of the basis is increased by letting $w_k = Av_k$. The vector w_k is then orthogonalized against $\{v_1, v_2, \dots, v_k\}$ by taking

$$w_k = w_k - \sum_{i=1}^{k} h_{ik}v_i, \quad v_{k+1} = \frac{w_k}{h_{k+1,k}}, \tag{2.1}$$

where $h_{k+1,k} = \|w_k\|_2$.

The coefficients h_{ik} in (2.1) form a $(k+1) \times k$ matrix, $\tilde{H}_k$. The jth column of $\tilde{H}_k$ contains the coefficients needed to orthonormalize v_{j+1} against the preceding basis vectors $\{v_1, v_2, \dots, v_j\}$ using the Gram-Schmidt process. We may choose a y that minimizes $\|\tilde{H}_k y - \beta e_1\|_2$, where βe_1 is a multiple of the first column of the $n \times n$ identity matrix and β is a constant. If V_k is a matrix which has the first k basis vectors of the Krylov subspace as its columns, the new approximate solution is $x_{k+1} = x_0 + V_k y$ as described in [**E**]. In exact arithmetic

$$\|b - Ax_k\|_2 = \|\tilde{H}_k y - \beta e_1\|_2. \tag{2.2}$$

In practice both sides of (2.2) are not equal due to loss of orthogonality of the basis vectors of K_m. The lack of orthogonality between basis vector may require reorthogonalization or the use of Householder transformations. The steps for constructing the $k+1st$ basis vector are outlined in Figure 1.

3. The Hessenberg Algorithm

Wilkinson presents Hessenberg's algorithm as an oblique projection [**H**] (pp.379-382). The following is a description of the algorithm without partial pivoting. Let e_i^T and e_j be the ith row and jth column respectively of an $n \times n$ identity matrix. Choose $l_1 = e_1$ and $y_1 = Al_1$. We can find y_2 by subtracting an appropriate multiple of l_1 from y_1 such that

$$y_2 = y_1 - a_{11}l_1 = \begin{pmatrix} 0 \\ a_{21} \\ a_{31} \\ \vdots \\ a_{n1} \end{pmatrix}$$

```
w_k = A v_k;                          /* enlarge subspace
For i = 1 : k,
    h_{i,k} = w_k^T v_i               /* modified Gram-Schmidt
    w_k = w_k - h_{i,k} v_i
End
h_{k+1,k} = ||w_k||_2
v_{k+1} = w_k / h_{k+1,k}             /* normalization step
```

FIGURE 1. **Formulation of the $k+1st$ basis vector using GMRES.**

and

$$l_2 = \frac{y_2}{a_{21}} = \begin{pmatrix} 0 \\ 1 \\ \tilde{a}_{31} \\ \vdots \\ \tilde{a}_{n1} \end{pmatrix}.$$

It is clear that $l_2 \perp e_1$. The first column of H is defined so that h_{11} is the multiple of l_1 needed to satisfy $l_2 \perp e_1$ and h_{21} is the scaling factor used to make $l_2(2) = 1$. This implies $h_{11} = a_{11}$ and $h_{21} = a_{21}$.

Now that we have l_2 and the first column of H, h_1, choose $y_1 = Al_2$. y_1 does not necessarily have zero entries. To insure $l_3 \perp e_1, e_2$ we must eliminate $y_1(1)$ and $y_1(2)$. This will be done by subtracting the appropriate multiple of l_1 and l_2 as follows:

$$\begin{aligned} y_1 &= Al_2 \\ y_2 &= Al_2 - \alpha_1 l_1 \\ y_3 &= Al_2 - (\alpha_1 l_1 + \alpha_2 l_2) = \begin{pmatrix} 0 \\ 0 \\ y_3(3) \\ \vdots \\ y_3(n) \end{pmatrix}. \end{aligned}$$

We choose constants α_1 and α_2 such that the proper multiple of l_1 and l_2 are subtracted from y_3, which insures the orthogonality of y_3 to e_1 and e_2. Now, $l_3 = y_3/\beta$ where $\beta = y_3(3)$. The second column of H, h_2, is created by letting $h_{12} = \alpha_1$, $h_{22} = \alpha_2$ and $h_{23} = \beta$.

This process is continued until at the kth step we obtain

$$\begin{aligned} y_1 &= Al_{k-1} \\ y_2 &= Al_{k-1} - \alpha_1 l_1 \\ &\vdots \\ y_{k-1} &= Al_{k-1} - (\alpha_1 l_1 + \alpha_2 l_2 + \ldots + \alpha_{k-2} l_{k-2}) \\ y_k &= Al_{k-1} - (\alpha_1 l_1 + \ldots + \alpha_{k-1} l_{k-1}). \end{aligned}$$

Here $\alpha_1 \ldots \alpha_{k-1}$ are the multiples of $l_1 \ldots l_{k-1}$ respectively, which eliminates elements of y_k to insure its orthogonality to $\{e_1, e_2, \ldots, e_{k-1}\}$. Substituting y_{k-1} into

```
w_k = A l_k                              /* enlarge subspace
For i = 1 : k,                           /* produces a zero in w_k(i)
    h_ik = w_k(i)
    w_k = w_k - h_ik l_i
End
h_{k+1,k} = w_k(k+1)
l_{k+1} = w_k / h_{k+1,k}                /* normalization step
```

FIGURE 2. **Formulation of the $k+1st$ basis vector using ELMRES without pivoting.**

the equation for y_k, it is easily seen that

$$y_k = y_{k-1} - \alpha_{k-1} l_{k-1} = \begin{pmatrix} 0 \\ \vdots \\ 0 \\ y_k(k) \\ \vdots \\ y_k(n) \end{pmatrix}.$$

The aforementioned algorithm can be described in the concise formula

$$y_{i-1} = A l_{k-1} - \sum_{j=1}^{i-1} h_{jk-1} l_j, \quad i = 2, \ldots, k \tag{3.1}$$

where h_{ik} is chosen so that

$$e_{i-1}^T y_i = e_{i-1}^T (y_{i-1} - h_{ik} l_i) = 0.$$

As a final step, the kth basis vector is constructed by $l_k = y_k / h_{k+1,k}$.

Wilkinson shows how to use partial pivoting for better stability. Let

$$\{e_1', e_2', \ldots e_{k-1}'\}$$

represent the permuted columns of an $n \times n$ identity matrix. The kth basis vector augments

$$span\ \{e_1', e_2', \ldots e_{k-1}'\}$$

with the k' position being chosen such that $h_{k+1,k}$ is maximized.

4. The ELMRES Algorithm

ELMRES uses the Hessenberg algorithm as described in Section 3 to construct a basis for a Krylov subspace. Let K_m be as in (1.2) with $l_1 = r/r(1)$ where $r = b - Ax$. [1] The procedure for extending the dimension of the subspace is shown in Figure 2.

Looking at the part the of algorithm which produces a new basis vector, we see that the ith step of the loop zeros $w_k(i), i = 1 \ldots k$. The normalization step forces the $k+1st$ entry of $l_{k+1} = 1$. Thus, $l_{k+1} \perp e_1, e_2, \ldots e_k$. When partial pivoting

[1] It should be noted that when pivoting is employed if $r(1) = 0$ we simply calculate the basis vectors in a permuted order say $1', 2', \ldots$. This is equivalent to solving the system $PAx = Pb$ where P is some permutation matrix.

is employed, the maximal entry of l_{k+1} in absolute value is one. Clearly $l_{k+1} \in span\{l_1, Al_1, \dots, A^k l_1\}$. Since the vectors $l_1, l_2, \dots l_{k+1}$ form a lower triangular matrix with ones on the diagonal, they are linearly independent and form a basis for the Krylov subspace. If l_1 is taken as e_1, the Hessenberg algorithm is equivalent to reducing the first columns of a matrix to upper Hessenberg form by elementary similarity transformations. The next few paragraphs explain the equivalence.

Recall that an elementary similarity transformation is of the form

$$L_i^{-1} = I - \hat{l}_i e_i^T, \quad L_i = I + \hat{l}_i e_i^T \tag{4.1}$$

where $l_i = [0 \dots 0 \quad 1 \quad \hat{l}_i]^T$ is a column vector with zeros as its first $i-1$ entries, 1 in the ith entry and a vector $\hat{l}_i$ of multipliers in the $i+1st$ to nth entries.

Neglecting pivoting, reduction by elementary similarity transformations to upper Hessenberg form is accomplished by

$$H = L_{n-1}^{-1} L_{n-2}^{-1} \dots L_2^{-1} A L_2 \dots L_{n-1}.$$

Therefore, the kth column of H is

$$H(:,k) = L_{n-1}^{-1} L_{n-2}^{-1} \dots L_2^{-1} A L_2 \dots L_{n-1} e_k.$$

Multiplying on the left by L_i^{-1} subtracts multiples of the ith row from the $i+1st$ to nth rows. Since the $k+2$ to n entries of $H(:,k)$ are zero

$$H(:,k) = L_{k+1}^{-1} L_k^{-1} \dots L_2^{-1} A L_2 \dots L_{n-1} e_k.$$

Multiplying on the right by L_i increments the ith column of H by $A\hat{l}_i$, leaving other columns of H unchanged. This implies

$$H(:,k) = L_{k+1}^{-1} L_k^{-1} \dots L_2^{-1} (A L_k e_k). \tag{4.2}$$

Grouping the operations as shown in (4.2) can reduce the overall number of operations because the multiplication

$$w_k = A L_k e_k = A l_k = A(:,k) + A(:,k+1:n)\hat{l}_k \tag{4.3}$$

uses the original sparse matrix A. Multiplications on the left by the L_i^{-1}'s are the saxpy operations $w_k = w_k - h_{ik} l_i$ of Figure 2.

Wilkinson [**H**] modified Hessenberg's method by introducing partial pivoting. In this case, $\|l_i\|_\infty = 1$. For ELMRES, pivoting is implicit in the sense that pivoting is performed by appropriate permutations of the basis vectors l_i, allowing matrix vector multiplications to be performed with the original matrix (See [**F**]). For sparse matrices, the first k Krylov basis vectors for the subspace

$$K = span\ \{r, Ar, \dots, A^{k-1} r\} \tag{4.4}$$

can be generated in $k^2 n - k^3/3$ flops. Getting a basis for the same subspace using GMRES requires $2k^2 n$ flops for modified Gram-Schmidt and $4k^2 n - (4/3)k^3$ flops for Householder transformations [**D**],[**G**].

L_m is a matrix where the first m columns are the basis vectors for the Krylov subspace and the last $n-m$ columns are the columns of an $n-m \times n-m$ identity matrix. It is written as follows:

$$
L_m = \left(\begin{array}{ccccc|cccc}
1 & 0 & 0 & \cdots & \cdots & 0 & \cdots & \cdots & 0 \\
\hat{l}_1 & 1 & \ddots & & . & & & & \vdots \\
\vdots & \hat{l}_2 & \ddots & \ddots & & & & & \vdots \\
\vdots & \vdots & \ddots & \ddots & \ddots & & & & \vdots \\
\vdots & \vdots & & \ddots & 1 & \ddots & & & \vdots \\
\vdots & \vdots & & & \hat{l}_m & 1 & \ddots & & \vdots \\
\vdots & \vdots & & & \vdots & 0 & \ddots & \ddots & \vdots \\
\vdots & \vdots & & & \vdots & \vdots & & \ddots & 0 \\
\vdots & \vdots & & & \vdots & 0 & \cdots & 0 & 1
\end{array} \right) = (\hat{L}_m | I_{n-m}) \tag{4.5}
$$

where $\hat{L}_m$ is the $n \times m$ matrix formed from the first m columns of L_m. The vectors (l_i) are constructed by the Hessenberg algorithm, such that

$$l_i \perp \{e_1, e_2, \dots, e_{i-1}\} \quad i = 2, \dots, m.$$

Define $r_0 = b - Ax_0$. Let $l_1 = [0, \dots, 0 \quad 1 \quad \hat{l}_1]^T$ such that

$$L_1^{-1} r_0 = (I - \hat{l}_1 e_1^T) r_0 = \beta e_1.$$

From Figure 2, we have

$$Al_k = \sum_{i=1}^{k+1} h_{ik} l_k \quad k = 1, \dots, m. \tag{4.6}$$

The matrix reformulation of (4.6) is

$$A\hat{L}_m = \hat{L}_{m+1} \tilde{H}_m$$

where $\hat{L}_{m+1}$ contains the first $m+1$ basis vectors for the Krylov subspace. We have constructed the bases $\hat{L}_m$ and $\hat{L}_{m+1}$ such that $\tilde{H}_m$ has $m+1$ rows, $m < n$ columns and is upper Hessenberg. We denote H_m as an $m \times m$ upper Hessenberg matrix which is equivalent to $\tilde{H}_m$ without its last row.

We want to choose the approximate solution

$$x_m = x_0 + \hat{L}_m y, \quad y \in R^m. \tag{4.7}$$

Multiplying both sides of the equation by A gives

$$Ax_m = Ax_0 + A\hat{L}_m y.$$

It follows that

$$
\begin{aligned}
b - Ax_m &= b - Ax_0 - A\hat{L}_m y \\
&= r_0 - A\hat{L}_m y.
\end{aligned}
$$

Let L_{m+1} be the matrix formed by augmenting the columns of $\hat{L}_{m+1}$ by the last $n-(m+1)$ columns of the $n \times n$ identity matrix. We now multiply by L_{m+1}^{-1} giving

$$
\begin{aligned}
L_{m+1}^{-1}(b - Ax_m) &= L_{m+1}^{-1}(r_0 - A\hat{L}_m y) \\
&= \beta e_1 - \left(\frac{H_m}{0 \quad \alpha e_1} \right) y
\end{aligned} \tag{4.8}
$$

where H_m is $m \times m$, the matrix $(0 \quad \alpha e_1)$ is $(n-m) \times m$, αe_1 is the $n-m$ vector $[\, \alpha\ 0\ \ldots\ 0\,]^T$, and $\alpha = h_{m+1,m}$.

Letting $\hat{\beta e_1}$ be the first m entries of βe_1 (the last $n-m$ entries of βe_1 are zeros), we choose y so that $H_m y = \hat{\beta e_1}$. With this choice,

$$\begin{aligned} L_{m+1}^{-1}(b - Ax_m) &= \left(\frac{0}{0 \quad \alpha e_1}\right) y \\ &= \left(\frac{0}{\alpha y(m) e_1}\right) \end{aligned}$$

where $\alpha y(m) e_1$ is an $n-m$ vector and $y(m)$ is the last entry of the m vector y . Multiplying on the left by L_{m+1} gives

$$\begin{aligned} b - Ax_m &= L_{m+1}\left(\frac{0}{\alpha e_1 y(m)}\right) \\ &= (\hat{L}_{m+1} | I_{n-m-1})\left(\frac{0}{\alpha e_1 y(m)}\right) \\ &= \left(\frac{0}{\alpha y(m) l_{m+1}}\right). \end{aligned} \tag{4.9}$$

Equation (4.9) is the Petrov-Galerkin conditions

$$\begin{aligned} b - A(x_0 + \hat{L}_m y) &\perp span\{e_i\} \quad i = 1, 2, \ldots, m \\ b - Ax_m &\perp span\{e_i\} \quad i = 1, 2, \ldots, m. \end{aligned}$$

We have chosen x_m in (4.7) by solving

$$H_m y = \hat{\beta e_1} \tag{4.10}$$

for an oblique projection method analogous to Full Orthogonalization Method (FOM). See for example (p. 124) [**D**].

Suppose we repartition (4.8) as

$$\begin{aligned} L_{m+1}^{-1}(b - Ax_m) &= L_{m+1}^{-1}(r_0 - A\hat{L}_m y) \\ &= \beta e_1 - \left(\frac{\tilde{H}_m}{0}\right) y \end{aligned} \tag{4.11}$$

and choose x_m as in (4.7) with y the least squares solution to $\|\tilde{H}_m y - \beta e_1\|_2$. $\tilde{H}_m$ is an $(m+1) \times m$ matrix. Let $z = \beta e_1 - \tilde{H}_m y$ be such that z is of minimal norm. Then

$$L_{m+1}^{-1}(b - Ax_m) = \left(\frac{z}{0}\right)$$

is minimal in 2-norm where

$$\|L_{m+1}^{-1}(b - Ax_m)\|_2 \tag{4.12}$$

is the elementary residual.

5. Numerical Results

We integrated a FORTRAN 77 version of ELMRES into the publicly available package SPARSKIT. [2] Using SPARSKIT enables us to compare ELMRES against GMRES. We are also able to test the behavior of ELMRES with restarts and using various preconditioners. The results quoted here were run on a 133 MHz Pentium

[2]Our only modification of SPARSKIT was the addition of one argument, an adjustable length integer vector to store pivot information. The ELMRES implementation uses the same backward communication organization as that used in the rest of the SPARSKIT package.

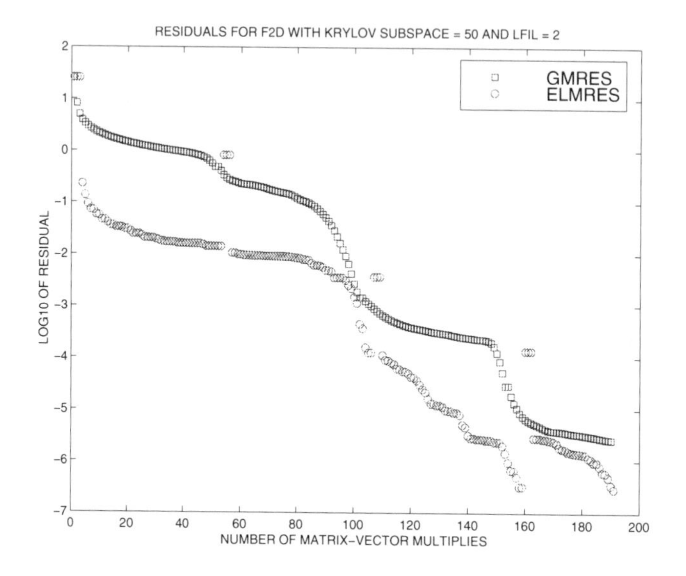

FIGURE 3. **F2D with lfil = 10, stopping tol = 10^{-7} and Krylov basis size 50.**

I with 48 MEG of RAM. We used Linux with the GNU g77 compiler and tuned BLAS [**A**].

Here we compare ELMRES to GMRES. Theoretically the GMRES residual satisfies $\|\bar{H}_m y - \beta e_1\|_2 = \|b - A\hat{x}\|_2$. For ELMRES, the computed residual on each step is $\|\bar{H}_m y - \beta e_1\|_2 = \|L_{m+1}^{-1}(b - A\hat{x})\|_2$. We explicitly calculate the mth residual, r_m, only after convergence is achieved in terms of $\|\bar{H}_m y - \beta e_1\|_2$. We then compute the ratio $\rho = \frac{\|\bar{H}_m y - \beta e_1\|_2}{\|A\hat{x} - b\|_2}$ and adjust the convergence tolerance by $\alpha = \frac{\alpha}{\rho}$. Achieving the adjusted tolerance often requires a few additional steps for ELMRES beyond the number required by GMRES.

We compared the number of iterations of GMRES and ELMRES with a variety of allowable levels of fill, $lfil$, for the preconditioner, ILUT. The parameter $lfil$ in the Incomplete LU factorization allow at most $lfil$ nonzero entries in a row of L and also $lfil$ nonzeros in a row of U. Entries of size less than 10^{-7} in L or U are dropped. For GMRES and ELMRES we also varied the size of the Krylov subspace accumulated before restarts.

The tolerance is calculated as in SPARSKIT [**C**]

$$residual = reltol\|b - Ax_0\|_2 + abstol. \tag{5.1}$$

The parameters *reltol* and *abstol* may be set by the user. It should be noted that for all of these experiments the solution is known. The right hand side of the system is generated by $A(1, \ldots, 1)^T = b$.

In our experiments, ELMRES and GMRES usually converge in almost the same number of iterations. If convergence is very fast, GMRES may be slightly faster because ELMRES performs an extra matrix vector multiply to pick out the first column of A. More often, ELMRES is significantly faster as it requires fewer operations and these execute more efficiently.

One example is the AF23560 matrix which is a 23560 by 23560 matrix from a discretization of the Navier-Stokes equation for an airfoil. It has 484256 nonzero entries. We used right preconditioning with ILUT having $lfil = 10$. Restarts were done every 70 steps. This matrix exhibits typical behavior for well behaved problems, with ELMRES requiring a few additional matrix vector multiplications and less time to converge (Table 2).

TABLE 2. **Times, number of matrix vector multiplications and residuals for AF23560 Matrix.**

	TIME (SEC)	MATVEC MULTIPLIES	RESIDUAL
GMRES	64.4 s	197	$1.062e-6$
ELMRES	46.4 s	209	$2.30e-6$

Another example is the F2D matrix which has 21904 by 21904 elements generated by discretizing the partial differential equation

$$\nabla^2 u + \frac{\partial u}{\partial x}\gamma e^{xy} + \frac{\partial u}{\partial y}\gamma e^{-xy} = h$$

on the rectangular grid $\Omega = (0,1)^2$ with Dirichlet boundary condition $u = 0$. In this experiment $lfil = 2$ and the Krylov basis size is set at 50. GMRES takes $26.8s$ to obtain a residual norm, $\|x-\hat{x}\|_2 = 2.52e-6$ while ELMRES obtains a residual norm of $\|x-\hat{x}\|_2 = 7.82e-6$ in $15.1s$. In Figure 5 we have plotted the $Log(residual)$ vs number of matrix vector multiplications where the residual is calculated by (5.1).

TABLE 3. **Comparison of times, matrix vector multiples, Av, and residuals for GMRES and ELMRES.**

GMRES			ELMRES		
TIME	**Av**	$\|\mathbf{x}-\hat{\mathbf{x}}\|_2$	**TIME**	**Av**	$\|\mathbf{x}-\hat{\mathbf{x}}\|_2$
ADD32, 23884 nonzeros, m = 30, tol = 1e − 2, lfil = 2					
$0.77s$	38	$1.1e-10$	$0.68s$	45	$2.7e-11$
FID10, 54816 nonzeros, m = 45, tol = 1e − 10, lfil = 60					
$7.0s$	137	$1.6e-8$	$4.8s$	97	$2.4e-8$
FID36, 53851 nonzeros, m = 50, tol = 1e − 7, lfil = 40					
$13.1s$	238	$7.3e-9$	$6.8s$	140	$6.1e-13$
FS541, 4285 nonzeros, m = 10, tol = 1e − 2, lfil = 2					
$0.01s$	7	$9.4e-10$	$0.01s$	9	$2.1e-11$
ORS1, 6858 nonzeros, m = 15, tol = 1e − 2, lfil = 5					
$0.26s$	37	$3.5e-6$	$0.27s$	43	$2.5e-6$
SAYLOR3, 375 nonzeros, m = 25, tol = 1e − 7, lfil = 10					
$0.03s$	7	$1.2e-6$	$0.03s$	9	$9.4e-8$
SHERMAN3, 20033 nonzeros, m = 35, tol = 1e − 7, lfil = 10					
$1.1s$	35	$9.3e-5$	$1.2s$	46	$4.9e-5$
UTM3940 , 83842 nonzeros, m = 65, tol = 1e − 10, lfil = 70					
$11.9s$	60	$3.7e-6$	$11.8s$	65	$5.5e-7$

In Table 3 the number of matrix vector multiplications, Av, and the residuals, $\|x-\hat{x}\|_2$ for GMRES and ELMRES are listed. Various matrices from the MATRIX MARKET collection [**B**] have been used for these experiments. The FID36 and FID10 matrices, from the SPARSKIT collection, are generated from finite element and finite difference schemes for PDE's. UTM5940 comes from the TOKOMAK collection which has matrices used in modeling Nuclear Physics. ORS1, SAYLR3 and SHERMAN3 come from oil reservoir problems. FS541 comes from the original set of Harwell-Boeing test matrices and represents one stage of FACSIMILE stiff ODE package for atmospheric pollution problems involving chemical kinetics and $2-D$ transport. All of the matrices are real-unsymmetric and poorly conditioned.

6. Conclusion

In this paper we have shown that the ELMRES algorithm using the Hessenberg method generates the first k basis vectors of the Krylov subspace in $k^2n - k^3/3$ flops. GMRES takes $2k^2n$ flops to form the same basis using modified Gram-Schmidt. We have also shown that while GMRES seeks an approximate solution

$$x_m = x_0 + V_m y \tag{6.1}$$

where y minimizes

$$\|\tilde{H}_m y - \beta e_1\|_2 = \|b - Ax\|_2$$

in exact arithmetic, ELMRES finds y such that the elementary residual,

$$\|L_{m+1}^{-1}(b - Ax)\|_2 = \|b - Ax\|_2$$

is minimized. Our numerical experiments have shown that ELMRES is as robust as GMRES and for problems requiring large basis size ELMRES converges faster than GMRES.

In the future we would like to exploit additional storage gained when performing computations in parallel. For GMRES, the number of messages to be passed per matrix vector multiplications is proportional to the basis size, while for ELMRES the number of messages passed is almost independent of Krylov basis size. Thus ELMRES should be relatively efficient in parallel. Robust convergence without good preconditioning is likely to be helpful in parallel computation.

References

[A] G. Henry. *Tuned blas library for Intel pentium processors running under Linux*, Technical report, University of Tennessee, Knoxville and Intel, 1998.

[B] R. Pozo and K. Remington. *Matrix market*, Technical report, http://math.nist.gov/MatrixMarket/, 1999.

[C] Y. Saad. *A basic tool kit for sparse matrix computations*, Technical Report 90-20, Research Institute for advanced Computer Science, NASA Ames research Center, 1990.

[D] Y. Saad. *Iterative Methods for Sparse Linear Systems*, ITP, PWS Publishing Company, Boston, 1996.

[E] Y. Saad and M. H. Schultz. *Gmres: A generalized minimal residual algorithm for solving nonsymmetric linear systems*, Siam J. Sci. Stat. Comp., 7:856–869, 1986.

[F] D. Stephens. *ELMRES: an Oblique Projection Method to Solve Sparse Non-Symmetric Linear Systems*, Ph.d. thesis, Florida Institute of Technology, Melbourne, FL, 1999.

[G] H. F. Walker. *Implementation of the GMRES method using Householder transformations*, SIAM J. Sci. Computing, 9:152–163, 1988.

[H] J. H. Wilkinson. *The Algebraic Eigenvalue Problem*, Clarendon Press, Oxford, England, 1965.

Department of Mathematics, Florida A & M University, Tallahassee, Florida 32307
E-mail address: dstephen@cis.famu.edu

Department of Mathematics, Florida Tech, Melbourne, Florida 32901
E-mail address: howell@zach.fit.edu

Contemporary Mathematics
Volume **275**, 2001

The Response of the Upper Ocean to Surface Buoyancy Forcing: A Characteristic Solution to Wave Propagation

Monica Y. Stephens and Zhengyu Liu

ABSTRACT. The evolution of the thermocline (the region of strong vertical temperature gradient that lies between the fairly homogeneous surface mixed layer and the deep ocean), within the northern hemisphere subtropical Pacific Ocean, is investigated using a two-layer planetary geostrophic model. This model aids in the understanding of the nonlinear response of the thermocline to local changes in surface heating/cooling. It is an idealized representation of the 2nd baroclinic mode (advective mode) in the interior of the thermocline and lends itself to solution by the method of characteristics.

The signal created by a heating/cooling anomaly propagates in a characteristic cone that follows the mean subsurface ventilation flow (i.e., flow that breaches the surface) which moves southwestward toward the equator. A linearization of the model equations reveals that the anomaly decays as it propagates through the subtropical ocean as a result of the divergence of the depth-averaged (barotropic) flow. The net divergence tends to spread the signal and increase its area, thereby reducing the amplitude of the signal.

It has been suggested that this advective mode provides a possible means of communication between the subtropical gyre and the tropical ocean on a decadal timescale, thus lending decadal variability to the tropical Pacific ocean. However, we find that the amplitude of the anomaly is reduced by almost half from its inception in the subtropical Pacific basin to its impact along the Asian coast. This suggests that some mechanism other than southward propagating subtropical anomalies may account for the bulk of tropical decadal variability.

1. Introduction

Ocean general circulation models (OGCMs) that simulate the full three-dimensional current, temperature, and salinity structure of the oceans are likely the most complete models available for studying and understanding the behavior of the oceans. However, more simple models may provide a better means of understanding certain aspects of ocean physics. In this paper we describe the use of a simple two-layer model of the upper ocean (upper 600 meters) to describe the propagation of mid-latitude ($\sim 35°N$ to $45°N$) temperature anomalies that originate off the coast of California and propagate southwestward towards the tropical Pacific ocean ($\sim 0°$

1991 *Mathematics Subject Classification.* Primary 86A05, 76U05; Secondary 76D33, 76B65.
Key words and phrases. Geophysical fluid dynanmics, oceanography.
The first author was supported by a post-doctoral fellowship at the UW-Madison.
The second author was supported by grants from NSF and NOAA.

to $25^\circ N$) on a decadal timescale. These anomalies are thought to provide a linkage between the mid-latitude (subtropical) ocean and the tropical ocean and thus lend decadal variability to the tropical oceans (Gu and Philander, 1997; Inui and Hanawa,1997). More recent analysis of observations of these temperature anomalies suggests that the anomalies undergo significant decay as they propagate and are therefore reduced to an extent that they have little impact on tropical variability (Deser et. al, 1996; Schneider et. al, 1999; Zhang and Liu, 1999). However, no theory that describes the mechanism for the decay of these anomalies presently exits.

Figure 1 (taken from Schneider et. al, 1999) shows the time evolution of two temperature anomalies from the early 1970s to the mid-1990s in the Pacific Ocean. There is a warm anomaly that originates at the surface at approximately $34^\circ N$ where it subducts and travels southward via the subsurface ocean towards the tropics to about $18^\circ N$. The warm anomaly is subsequently followed by the subduction of a cold temperature anomaly within the same region. The amplitude of the warm anomaly is significantly reduced during its propagation. The cold anomaly also initially undergoes some decay but later amplifies. However, the amplification of the cold anomaly is a result of local tropical wind forcing and is not a result of the subduction process itself (Schneider et. al, 1999). One might assume that these subduction temperature (active tracer) anomalies propagate similar to a passive tracer in the flow, being advected along by the subsurface or ventilation flow field and dissipated by turbulent mixing alone. By definition, a passive tracer, unlike an active tracer, is a benign substance that enters the ocean without altering the flow characteristics of the water. An active such as a temperature (height) anomaly affects the density of the ocean waters and therefore affects the flow. Although turbulent dissipation does play a role in the decay of active tracer anomalies, a significant portion of the decay is related to wave dynamics.

This paper presents a simple explanation for the dissipation of these active tracer anomalies, owing to planetary wave dynamics, using a two-layer model of the ocean. We first discuss the model development and physics and the solution of the model in an unperturbed state in which there is no anomalous forcing, particularly the setup of the ventilated region of flow. Next we discuss the perturbed solution in which we add buoyancy forcing by altering the position of the outcrop line similar to Huang and Pedlosky (1999). We present a mechanism for the decay of these anomalies and compare the propagation of these anomalies (active tracer) to the passage of a passive tracer within the basin. Lastly, we present our summary and conclusions.

2. Two-Layer Planetary Geostrophic Model

The two-layer planetary geostrophic model of the ventilated thermocline provides a dynamical framework for investigating the amplitude of anomalies that are subducted in the subtropical Pacific basin. A three-dimensional representation of the model is shown in Figure 2.

The upper ocean has a relatively thin mixed-layer (upper 100 meters) with fairly homogeneous temperature and salinity that remains well-mixed because of the action of the winds at the surface. The mixed-layer is underlain by a well-stratified temperature layer called the thermocline. The layers of the two-layer model represent the stratification within the thermocline with the lower layer being

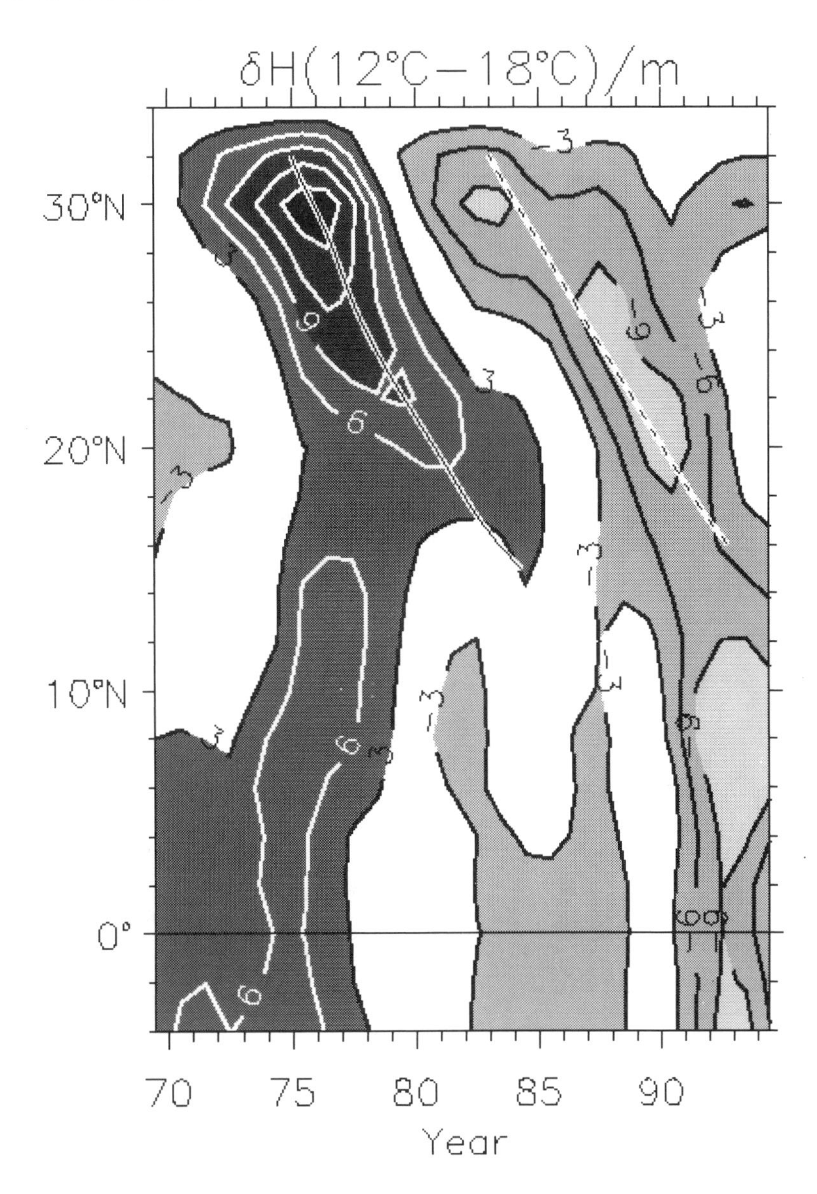

FIGURE 1. Observations of decadal temperature anomalies in the Pacific basin (degrees C). Zonal average of decadal anomalies between the depth of the $12°C$ and $18°C$ isotherms. A warm temperature anomaly is followed by the subduction of a cold temperature anomaly between 1969 and 1993. There is evidence of amplitude decay for both anomalies.(Schneider et. al, 1999)

colder than the upper layer. The model domain consists of the subtropical Pacific basin from $12°N$ to $40°N$. The longitudinal extent of the model basin is $L =$ 8400 kilometers with the eastern and western boundaries specified by $x_e = 0$ and $x_w = -1$, respectively. The lower layer outcrops along a zonal latitude circle at

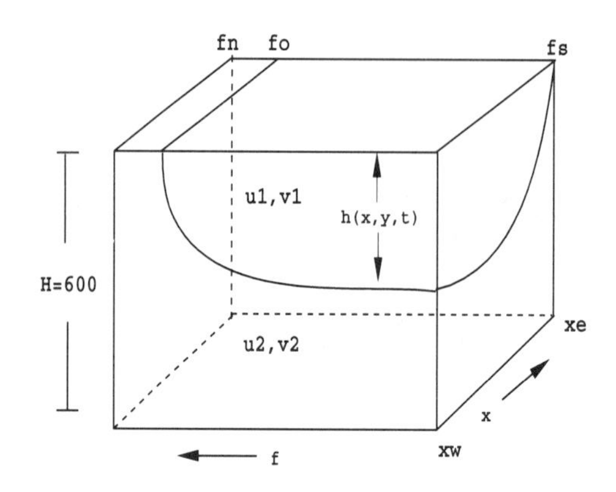

FIGURE 2. 3-D figure of the two-layer planetary goestrophic model. The interface between the two layers is specified by h. The base of the model (base of the thermocline) is fixed at $H = 600$ meters. The lower layer outcrops along a zonal outcrop line. The eastern and western boundaries of the model are specified by $x_e = 0$ and $x_w = -1$ with a longitudinal extent of $L = 8400$ kilometers. The latitudinal extent of the basin is from $12°N$ to $40°N$.

$36.5°N$ in the northern portion of the basin. This latitude is represented by the corresponding value of the Coriolis parameter, $f_o = 2\Omega sin(36.5)$. The base of the model represents the base of the thermocline with a fixed depth of, H = 600 meters. A fixed thermocline base is not strictly true in the real ocean. However, this simplification filters out the lower modes, such as the first-baroclinic ocean mode, that travel due west of the disturbance region and therefore do not propagate toward the tropical oceans (Liu, 1999). We therefore preferentially isolate signals of the second-baroclinic type, also known as the subduction or advective mode, that propagate southwestward, towards the tropics, with the subduction flow field.

The waters within the model basin are driven by surface wind stress and buoyancy-induced anomalies in the model are generated by a perturbation of the zonal outcropping latitude. From Ekman theory the negative windstress curl in the subtropical Pacific basin, from a combination of the low latitude trade winds and the mid-latitude westerlies, causes a net downward velocity at the top of the thermocline (bottom of the mixed layer) called Ekman pumping ($w_e < 0$). This downward Ekman pumping velocity is responsible for the subduction of the cold subtropical waters beneath the warmer tropical waters.

The model physics is the same as described in Liu (1993a,b). Each of the model layers is in geostrophic balance:

$$u_1 = -\frac{g}{f}\frac{\partial \eta}{\partial y}, \qquad v_1 = \frac{g}{f}\frac{\partial \eta}{\partial x}, \tag{2.1}$$

$$u_2 = -\frac{1}{f}\left(g\frac{\partial \eta}{\partial y} + \gamma\frac{\partial h}{\partial y}\right), \qquad v_2 = \frac{1}{f}\left(g\frac{\partial \eta}{\partial x} + \gamma\frac{\partial h}{\partial x}\right), \tag{2.2}$$

where $\vec{v_1} = (u_1, v_1)$ and $\vec{v_2} = (u_2, v_2)$ are the velocities in layers 1 and 2, h is the depth of the interface separating the layers, η is the surface elevation, $g = 9.8ms^{-2}$

is the gravity, and $\gamma = g\Delta\rho/\rho_o \approx 2cms^{-2}$ is the reduced gravity. The difference in density between layers 1 and 2, $\Delta\rho$, is much less than the mean density, $\rho_o = 1000kgm^{-3}$.

The model interior flow is controlled by Sverdrup dynamics which is the vorticity form of the geostrophic balance, assuming the Coriolis parameter, f, is linear a function of latitude. Thus,

$$\beta v_B = f w_e, \tag{2.3}$$

where $\vec{v}_B = (u_B, v_B)$ is the barotropic or depth-averaged flow field with $\vec{v}_B = [\vec{v}_1 h + \vec{v}_2(H-h)]/H$, $\beta = df/dy = (2\Omega/R)\, cos(35^\circ)$ is the mean β-parameter for the basin and $R = 6.37 \times 10^6 m$ is the mean radius of the earth, and $w_e = -H(\nabla \cdot \vec{v_B})$ is the Ekman pumping velocity. Additionally, the waters within the lower layer conserve potential vorticity once they are subducted beneath the upper layer. Therefore, given the potential vorticity, $q = f/(H-h)$,

$$\left(\frac{\partial}{\partial t} + \vec{v_2}\right) q = 0. \tag{2.4}$$

For convenience, the Sverdrup relation (Eq. 2.3) is zonally integrated from the eastern boundary ($x_e = 0$) to the ocean interior yielding

$$\left[2H\eta + (H-h)^2\right] = \left[2H\eta_e + (H-h_e)^2\right] + 2f^2 \int_o^x w_e(x,f,t)dx/\beta\gamma. \tag{2.5}$$

η_e and h_e are the surface elevation and interface depth which are taken to be zero to satisfy no flow at the eastern wall (California coast). Equation (2.5) is combined with equation (2.4) to derive an equation that describes the evolution of the interface separating the layers which is given by,

$$\frac{\partial h}{\partial t} = \left[u_B + c(h)\right]\frac{\partial h}{\partial x} + \beta v_B \frac{\partial h}{\partial f} = -(1 - h/H)w_e, \tag{2.6}$$

where $f = f_m + \beta y$, with $f_m = 2\Omega sin(35)$, has been used as the meridional coordinate in lieu of y since f varies monotonically with y. The barotropic (depth-averaged) velocity vector, $\vec{v}_B = [\vec{v}_1 h + \vec{v}_2(H-h)]/H$, has components

$$(u_B, v_B) = \left(-\frac{\partial_f\left[\rho_o \gamma w_e(f)x\right]}{2Hf\rho_o}, \frac{fw_e}{\beta H}\right). \tag{2.7}$$

The Rossby wave speed $c(h)$ is given by,

$$c(h) = -\beta\gamma h(H-h)/f^2 H. \tag{2.8}$$

Equation (2.6) contains two boundary conditions; one along the outcrop line that defines the ventilated region of the model, and the other along the eastern boundary that defines the unventilated or shadow region of the model which is just west of the eastern boundary and contains no flow. The boundary conditions are as follows:

$$\begin{aligned} h(x, f_o, t) &= 0 \\ h(x_e, f, t) &= 0. \end{aligned} \tag{2.9}$$

Since the lower layer breaches the surface along the outcrop line, the interface depth is zero. Along the eastern boundary the interface depth is constant and taken to be zero to satisfy a no zonal flow boundary condition.

The model is forced by Ekman pumping that is zonally uniform and independent of time:

$$w_e(f) = -w_o(f^2 - (1 + f_s)f + f_s), \tag{2.10}$$

where $w_o = 6 \times 10^{-4} cms^{-1}$ so that $w_e = 10^{-4} cms^{-1}$ is the maximum strength of the Ekman pumping rate. The magnitude of the Ekman pumping is maximal at the center of the basin and reduces to zero at the northern ($f_n = 2\Omega sin(40°)$) and southern ($f_s = 2\Omega sin(12°)$) boundaries of the basin, reminiscent of the net convergence of the Ekman transport that leads to Ekman pumping ($w_e < 0$) in the subtropical Pacific basin.

Buoyancy-induced anomalies in the model are forced by a perturbation of the zonal outcrop line. A warm (cold) anomaly results from a northward (southward) displacement of the outcrop line. This anomalous forcing is specified by,

$$f_o(x,t) = f_c + ae^{(x-x_o)^2/\alpha^2} sin(\omega t), \tag{2.11}$$

where $a = 1.125°$ latitude is the amplitude of the anomalous outcropping latitude, $x_o = -0.3/L$ is the longitudinal location of the center of the anomaly, $\alpha = 84$ km is the e-folding distance for the zonal extent of the anomaly, $\omega = 2\pi/20yrs$ is its forcing frequency, and $f_c = 2\Omega sin(36.5)$ is the mean position of the outcrop line. A perturbation of the outcrop line by $a = 1.125°$ latitude results in a temperature anomaly of approximately $1.125°C$ (in winter) and an amplitude of approximately 15 meters. Physically, these anomalies may be created by some warm (cold) air mass that advects over the ocean creating a region of localized heating (cooling).

In addition to the buoyancy-induced anomalies, a passive tracer is also introduced into the model to compare its amplitude evolution to that of the buoyancy anomaly or active tracer. The passive tracer, whose amplitude is given by

$$Tr = (1)e^{(x-x_o)^2/\alpha^2} sin(\omega t), \tag{2.12}$$

is released in the same location as the active tracer. The passive tracer is advected along by the lower layer (ventilation) flow field. Its amplitude evolution is governed by

$$\frac{\partial T_r}{\partial t} + u_2\frac{\partial T_r}{\partial x} + v_2\frac{\partial T_r}{\partial f} = 0, \tag{2.13}$$

where (u_2, v_2) are the velocities in the subsurface layer.

3. The characteristic solution

Equation (2.6) lends itself to solution by the method of characteristics. We adopt characteristic variable s and obtain a set of characteristic curves

$$\begin{aligned} (a)\frac{dt}{ds} &= 1 \\ (b)\frac{dx}{ds} &= u_B + c(h) \\ (c)\frac{df}{ds} &= \beta v_B \\ (d)\frac{dh}{ds} &= -(1 - h/H)w_e \end{aligned} \tag{3.1}$$

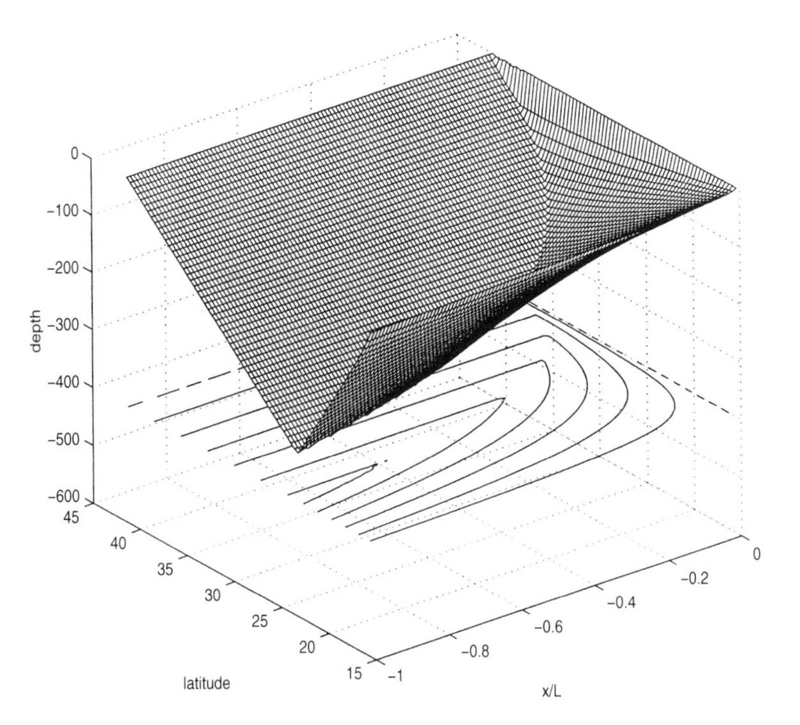

FIGURE 3. Model-derived interface depth. The maximum depth is 300 meters along the western boundary. The contours are in 50 meter intervals.

which are solved in both the ventilated region, defined by the characteristic lines that emanate from the outcrop line, and the shadow region, defined by the characteristics that emanate from the eastern boundary. The boundary conditions for the interface equation serve as the initial conditions for the characteristic curves in both regions. In the ventilated zone $s = 0$ corresponds to $h_i = 0$ and $f_i = f_o$. In the shadow zone $s = 0$ corresponds to $h_i = 0$ and $x_i = x_e = 0$. In this study, we are specifically interested in the propagation of Rossby waves in the ventilated region of the model. Therefore, only the solution in the ventilated region will be shown. The solution in the shadow region is given in Liu (1993a).

The solution in the ventilated zone is found by dividing (3.1c) by (3.1d) which yields

$$h = H(1 - f/f_i) = H(1 - f/f_o). \tag{3.2}$$

This result is simply a restatement of the conservation of potential vorticity. The depth of the interface decreases to the south, in the basin, with a decrease in the Coriolis parameter, f.

Figure 3 shows the solution for the interface for the unperturbed case ($a = 0$). The interface is bowl-shaped and reaches a maximum depth of 300 meters along the western boundary. This area of maximum depth is formed by flow that originates in the ventilated region of the basin. To the south of the maximum, the depth of the interface shoals. This shoaled region is formed by characteristics that originate along the eastern boundary and are within the shadow or no flow region of the basin.

The active tracer anomaly from the perturbed solution ($a = 1.125$) is now compared to the propagation of a passive tracer in the model. The amplitude of the active tracer anomalies is obtained by taking the difference in the perturbed solution ($a = 1.125$) from the unperturbed solution ($a = 0$). Figure 4 shows snapshots, at five year intervals, of the height (active tracer) anomaly and the passive tracer anomaly. The anomalies subduct at $36.5^\circ N$ and travel southwestward in the ventilated region until the they reach the western boundary at approximately $22^\circ N$. In the absence of any turbulent decay mechanism, the passive tracer anomaly preserves its amplitude during its propagation. The amplitude of the negative passive tracer anomaly (dashed lines) is identical to its amplitude at the western boundary. However, the active tracer anomaly decays from a maximum amplitude of 16 meters to approximately 8 meters along the western boundary. This indicates that the amplitude decay of the active tracer anomaly is related to planetary wave dynamics. In the real ocean, a passive tracer anomaly would undergo some decay due to the presence of turbulent mixing. An active tracer anomaly would experience decay from both planetary wave dynamics and turbulent mixing. This simple two-layer model indicates that the active tracer anomaly decays by half due to planetary wave dynamics. We now explore the actual physical mechanism that leads to the amplitude decay of the active tracer in this two-layer example.

4. Discussion

Liu (1993b) used a linear perturbation analysis to explain the decay of a height anomaly caused by anomalous, time-dependent Ekman pumping. We use the same result here to describe the amplitude evolution of the active tracer anomalies. The interface equation (Eq. 2.6) is linearized with an overbar representing mean quantities and a prime representing first-order perturbations from the mean. The anomaly equation is therefore

$$\frac{\partial h'}{\partial t} + \bar{\vec{v_B}} \cdot \nabla h' + c(\bar{h})\frac{\partial h'}{\partial x} = -\mu h', \tag{4.1}$$

where $\mu = \nabla \cdot \bar{\vec{v_B}}$ is the horizontal divergence of the barotropic (depth-averaged) flow. This equation is reminiscent of the mass conservation equation with an additional nonlinear term $c(\bar{h})\partial h'/\partial x$ that describes the advection by the Rossby wave speed. Therefore, the total rate of change of the amplitude of the height anomaly is determined by the divergence of the mean barotropic flow. The divergence is related to the Ekman pumping velocity $-\nabla \cdot \bar{\vec{v_B}} = w_e/H$. Since $w_e < 0$, μ is positive, and the anomaly tends to decay as it travels through the basin. Physically speaking, the mean divergence expands the area, A, of the anomaly, thereby decreasing its amplitude. Given $(1/A)dA/dt = \nabla \cdot \bar{\vec{v_B}} = -we/H$, then

$$\frac{\partial h'}{\partial t} + \bar{\vec{v_B}} \cdot \nabla h' + c(\bar{h})\frac{\partial h'}{\partial x} = -\frac{1}{A}\frac{dA}{dt}. \tag{4.2}$$

Therefore, the active tracer anomaly, according to this two-layer model result, decays as a result of the net divergence of the mean barotropic flow field in the subtropical Pacific basin. This divergent flow field, caused by the negative Ekman pumping in the basin, spreads the area of the anomaly and therefore decreases its amplitude. In comparison, the passive tracer, given by equation (2.13) is simply advected by the mean subsurface or ventilation flow field without experiencing any change in its amplitude.

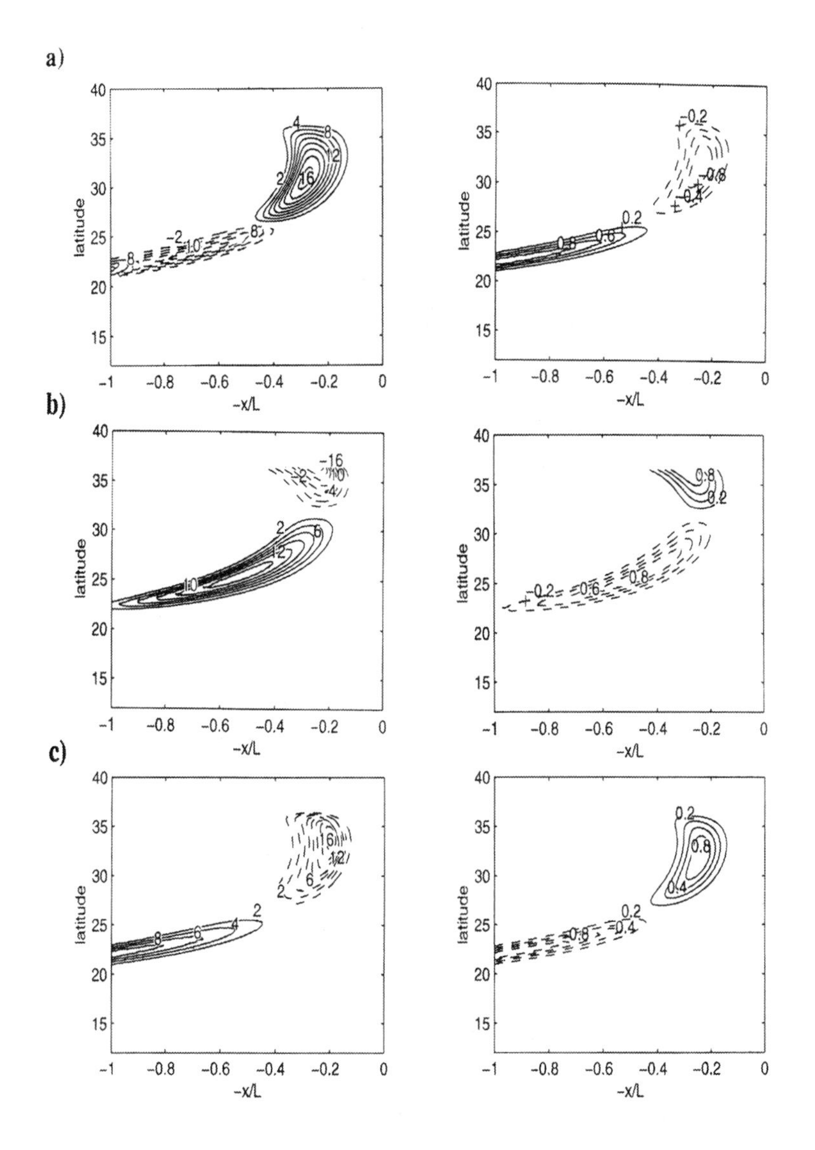

FIGURE 4. Snapshots of the active tracer anomaly and passive tracer anomaly at five year increments (a) t_o, (b) $t_o + 5$ yrs, and (c) $t_o + 10$ yrs. The progression of the active tracer anomaly is shown in the left column and the progression of the passive tracer anomaly is shown in the right column.

5. Summary and Conclusions

We have developed a theory to describe the amplitude evolution of thermocline anomalies that subduct and propagate in the subtropical Pacific basin using a simple two-layer model of the ocean thermocline. The amplitude of these active tracer anomalies is found to differ significantly from a passive tracer because of planetary wave dynamics.

The amplitude of an active tracer (temperature or height anomaly) decays considerably more than a passive tracer. The amplitude of the active tracer decays as a result of two primary mechanisms; (1) turbulent diffusion and (2) the divergence of the mean barotropic flow in the subtropical Pacific basin. In a two-layer model of the ocean, in which no turbulent decay mechanism has been introduced, we find that the divergence of the mean barotropic flow field causes the active tracer anomaly to decay by approximately half its original amplitude. A passive tracer, in the same model, shows no significant decay. In the real ocean, where turbulent diffusion is present, both the active and passive tracers would experience decay. However the additional decay due to planetary wave dynamics would cause the active tracer to decay more significantly than the passive tracer. A numerical experiment using a complex three-dimensional ocean general circulation model (OGCM) reveals that the amplitude of the active tracer anomaly is almost completely depleted prior to reaching the tropical oceans which suggests that these active tracer anomalies (temperature anomalies) have a minimal impact on the tropical ocean temperatures. A more general result describing both the amplitude and speed of these thermocline anomalies can be found in Stephens et. al (2000).

One note of interest is that the waters within ocean gyres are not always divergent which may lead to a result different from the one we have obtained here. For example, in the high latitude oceans (greater than $55°N$) a subpolar gyre exists in the Pacific and Atlantic oceans that has surface Ekman suction ($w_e > 0$) and net convergent flow which could possibly lead to an amplification of subduction anomalies.

References

[A] Deser, C., M.A. Alexander and M.S. Timlin, *Upper ocean thermal variations in the North Pacific during 1970-1991*, J. Climate, vol. **9**, 1996, pp. 1840–1855.

[B] Gu, D., and G. S. Philander, *Interdecadal climate fluctuations that depend on exchanges between the tropics and the extratropics*, Science, vol. **275**, 1997, pp. 805–807.

[C] Huang, R.X. and J. Pedlosky, *Climate variability inferred from a layered model of the ventilated thermocline*, J. Phys. Oceanogr., vol. **29**, 1999, pp. 779–790.

[D] Inui, T. and K. Hanawa, *A numerical investigation of effects of a tilt of the zero wind stress curl line on the subduction process*, J. Phys. Oceanogr., vol. **27**, 1997, pp. 897–908.

[E] Liu, Z., *Thermocline forced by Ekman pumping. I: spinup and spindown*, J. Phys. Oceanogr., vol. **23**, 1993a, pp. 2505–2522.

[F] Liu, Z., *Thermocline forced by varying wind. II: annual and decadal Ekman pumping*, J. Phys. Oceanogr., vol. **23**, 1993b, pp. 2523–2541.

[G] Schneider, N., A. J. Miller, M.A. Alexander and C. Deser, *Subduction of decadal North Pacific temperature anomalies: observations and dynamics*, J. Phys. Oceanogr., vol. **29**, 1999, pp. 1056–1070.

[H] Stephens, M.Y., Z. Liu, and H. Yang, *On the evolution of subduction planetary waves with application to north Pacific decadal climate variability*, J. Phys. Oceanogr., 2000 (in press).

[I] Zhang R. H. and Z. Liu, *Decadal thermocline variability in the North Pacific: Two pathways around the subtropical gyre*, J. Climate, vol. **12**, 1999, pp. 3273–3296.

Department of Atmospheric and Oceanic Sciences, University of Wisconsin-Madison, Madison, Wisconsin 32312

Current address: Department of Oceanography, Florida State University, Tallahassee, Florida 43403

E-mail address: mys@rossby.ocean.fsu.edu

Department of Atmospheric and Oceanic Sciences, University of Wisconsin-Madison, Madison, Wisconsin 32312

E-mail address: zliu@facstaff.uwm.edu

Contemporary Mathematics
Volume **275**, 2001

An Overview of Integro-Differential Equations and Variational Lyapunov Method

Sonya A. F. Stephens and V. Lakshmikantham

ABSTRACT. Much has been done to unify the method of variation of parameters and Lyapunov's second method to analyze the stability characteristics of a perturbed system. This method, the variational Lyapunov method, is applied to integro-differential equations. Here we investigate a nonlinear variation of parameters along with a comparison result which will allow us to examine the stability of the trivial solutions of perturbed systems.

1. Introduction

Recently, by unifying the method of variation of parameters [**A**] and the Lyapunov method [**C, D**], a fruitful technique was developed for differential equations which is called the variational Lyapunov method. The advantage of this unification is shown in [**B, D**] by integrating stability criteria. This paper will exploit this idea to study the stability properties of solutions of integro-differential equations of Volterra type. We shall first develop the nonlinear variation of parameters formula from a slightly different point of view to show its advantage. Then we proceed to develop a general comparison theorem which incorporates both the method of variation of parameters and Lyapunov method. Since the estimation of the derivative of a Lyapunov function relative to the integro-differential system requires the selection of a minimal class of functions, the comparison theorem is proved in the most general framework to include several known cases. Finally we present a brief example from [**D**] that gives an overview of the ideas presented in this paper.

2. Nonlinear Variation of Parameters

Consider the differential system

$$y' = F(t, y), \quad y(t_0) = x_0, \tag{2.1}$$

and the integro-differential system

$$x' = f(t, x, Tx), \quad x(t_0) = x_0, \tag{2.2}$$

where $f \in C[R_+ \times R^n \times R^n, R^n], F \in C[R_+ \times R^n, R^n]$, and $(Tx)(t) = \int_{t_0}^{t} K(t, s, x(s))ds$ where $K \in C[R_+^2 \times R^n, R^n]$. Let $y(t) = y(t, t_0, x_0)$ be the unique solution of (2.1)

1991 *Mathematics Subject Classification.* Primary 34D05,34D10,34D20; Secondary 45J05.
Key words and phrases. Integro-Differential Equations, Variation of parameters, Stability.

existing for $t \geq t_0$ such that $\frac{\partial y}{\partial t_0}(t, t_0, x_0)$ and $\frac{\partial y}{\partial x_0}(t, t_0, x_0)$ exist and are continuous on $R_+ \times R^n$. Then setting $v(s) = y(t, s, x(s))$, $t_0 \leq s \leq t$, where $x(s) = x(s, t_0, x_0)$ is any solution of (2.2) existing on $[t_0, \infty)$, we get

$$\begin{aligned} \frac{dv(s)}{ds} &= \frac{\partial y}{\partial t_0}(t, s, x(s)) + \frac{\partial y}{\partial x_0}(t, s, x(s)) f(s, x(s), (Tx)(s)) \\ &\equiv G(t, s, x(s), (Tx)(s)). \end{aligned} \tag{2.3}$$

Then integrating from t_0 to t, we have

$$x(t) = y(t) + \int_{t_0}^{t} G(t, s, x(s), (Tx)(s)) ds, \quad t \geq t_0, \tag{2.4}$$

which is the nonlinear variation of parameters formula. To see this, let

$$f(t, x, Tx) = F(t, y) + R(t, Tx) \tag{2.5}$$

and suppose that $\frac{\partial F}{\partial y}(t, y)$ exists and is continuous on $R_+ \times R^n$. Then we know that $\frac{\partial y}{\partial t_0}(t, t_0, x_0)$ and $\frac{\partial y}{\partial x_0}(t, t_0, x_0)$ are the solutions of the variational system

$$z' = F_y(t, y(t)) z,$$

satisfying the initial data

$$\begin{aligned} \frac{\partial y}{\partial t_0}(t_0, t_0, x_0) &= -F(t_0, x_0), \\ \frac{\partial y}{\partial x_0}(t_0, t_0, x_0) &= I \end{aligned}$$

where I is the identity matrix and the following identity is true

$$\frac{\partial y}{\partial t_0}(t, t_0, x_0) + \frac{\partial y}{\partial x_0}(t, t_0, x_0) F(t_0, x_0) \equiv 0. \tag{2.6}$$

Consequently, we obtain from (2.3),(2.4),(2.5), and (2.6), the relation

$$x(t) = y(t) + \int_{t_0}^{t} \frac{\partial y}{\partial x_0}(t, s, x(s)) R(s, (Tx)(s)) ds, \tag{2.7}$$

which is analogous to Alekseev's nonlinear variation of parameters formula for differential equations, see [**C**]. When we utilize (2.7) or (2.4) to derive qualitative properties of the solution of (2.2), it will allow us to estimate $|x(t)|$ which leads to estimating $|\frac{\partial y}{\partial x_0}(t, t_0, x_0)|$ and $|R(s, (Tx)(s))|$ which in fact destroys any good behavior of the perturbation term R.

If on the other hand, we proceed in a different way, by setting $v(s) = |y(t, s, x(s))|^2$, then we would obtain, proceeding as before, the relation for $t \geq t_0$,

$$|x(t)|^2 = |y(t)|^2 + \int_{t_0}^{t} 2y(t, s, x(s)) G(t, s, x(s), (Tx)(s)) ds. \tag{2.8}$$

This suggests that we can employ in a more general manner a function $V \in C'[R_+ \times R^n, R_+]$ so that one obtains the relation

$$V(t,x(t)) = V(t_0, y(t)) + \int_{t_0}^{t} \frac{\partial V}{\partial s}(s, y(t,s,x(s)))ds$$

where

$$\frac{\partial V}{\partial x}(s,x(s)) \equiv \frac{\partial V}{\partial t}(t,x) + \frac{\partial V}{\partial x}(t,x)G(t,s,x(s),(Tx)(s)). \tag{2.9}$$

Clearly, (2.8) or (2.9) have an advantage over (2.4). Hence as it is, we can extract any good behavior of this integral term, if any. What is done here is the idea of the unification of variation of parameters and the Lyapunov method so that one can reap the advantage of both methods. This is precisely what we do in the next section in a more general way.

3. Nonlinear Variation of Parameters

Considering equations (2.1) and (2.2) with previously mentioned conditions and solution representations, we define for $t_0 \le s \le t$

$$\begin{aligned} D\_V(s,y(t,s,x(s))) \quad &\equiv \quad \lim_{h\to 0^-} \frac{1}{h}[V(s+h, \\ &\quad x(s)+hf(s,x(s),(Tx)(s))) - V(s,y(t,s,x(s)))]. \end{aligned}$$

where $V \in C[R_+ \times R^n, R_+]$ and $x \in C[R_+, R^n]$. We then have the following general comparison theorem.

THEOREM 3.1. *Assume that*

1. $V \in C[R_+ \times R^n, R_+]$, $V(t,x)$, *and* $|y(t,s,x)|$ *are locally Lipschitzian in* x;
2. $g_0(t,s,u) \le g(t,s,u)$, g_0, $g \in C[R_+^2 \times R, R]$, $\eta(t,s,t^0,v_0)$ *is the left maximal solution of*

$$\frac{\partial v(s)}{\partial s} = g_0(t,s,v(s)),\ v(t_0) = u_0 \ge 0, \tag{3.1}$$

   exists for $t_0 \le s \le t^0 \le t$, *and* $r(t,s,t_0,u_0)$ *is the right maximal solution of*

$$\frac{du}{ds} = g(t,s,u),\ u(t_0) = v_0 \ge 0, \tag{3.2}$$

   existing on $t_0 \le s \le t < \infty$;
3. $\Omega = [x \in C[R_+, R^n] : V(s,y(t,s,x(s))) \le \eta(t,s,s_1,V(s_1,y(t,s_1,x(s_1))))$ *for* $t_0 \le s \le s_1 \le t]$;
4. *for* $t_0 \le s \le t < \infty$, $x(s) \in \Omega$,

$$D\_V(s,y(t,s,x(s))) \le g(t,s,V(s,y(t,s,x(s)))).$$

Then any solution, $x(t,t_0,x_0)$, *satisfies the estimate*

$$V(t,x(t,t_0,x_0)) \le r_0(t,t_0,V(t_0,y(t,t_0,x_0))), t \ge t_0, \tag{3.3}$$

where $r_0(t, t_0, u_0) = r(t, t, t_0, u_0)$.

We need the following lemma to prove this result [**F**].

LEMMA 3.2. *Assume assumption (2) from Theorem 3.1 holds. Then whenever* $r(t, t^0, t_0, u_0) \leq v_0$, *we have*

$$r(t, s, t_0, u_0) \leq \eta(t, s, t^0, v_0)$$

for $t_0 \leq s \leq t^0 \leq t$.

Now we shall prove Theorem 3.1.

PROOF. Set $m(t, s) = V(s, y(t, s, x(s)))$ where $x(t) = x(t, t_0, x_0)$ is any solution of (2.2) existing on $[t_0, \infty)$. Then $m(t, t_0) = V(t_0, y(t, t_0, x_0)) \leq u_0$. We shall first show that

$$m(t, s) < u(t, s, \epsilon), \; t_0 \leq s \leq t \tag{3.4}$$

where $u(t, s, \epsilon)$ is any solution of

$$\frac{du}{ds}(t, s, \epsilon) = g(t, s, u(s)) + \epsilon, \; u(t_0) = u_0 + \epsilon,$$

which exists as far as the maximal solution $r(t, s, t_0, u_0)$ exists and satisfies $\lim_{\epsilon \to 0} u(t, s, \epsilon) = r(t, s, t_0, u_0)$ for $t_0 \leq s \leq t < \infty$. If (3.4) is not true, then there exists a s_1 with $t_0 \leq s \leq s_1 \leq t$ such that

$$m(t, s_1) = u(t, s_1, \epsilon)$$

and

$$m(t, s) < u(t, s, \epsilon)$$

$t_0 \leq s < s_1 \leq t$. We then get

$$D_{-}m(t, s_1) \geq \frac{du}{ds_1}(t, s_1) = g(t, s_1, u(s_1)) + \epsilon. \tag{3.5}$$

Now consider the left minimal solution, $\eta(t, s, s_1, m(t, s_1))$ of (3.4) for $t_0 \leq s \leq s_1 \leq t$. By Lemma 3.2, we have

$$r(t, s, t_0, u_0) \leq \eta(t, s, s_1, m(t, s_1)), \quad t_0 \leq s \leq s_1 \leq t,$$

since $\lim_{\epsilon \to 0} u(t, s_1, \epsilon) = r(t, s_1, t_0, u_0)$ and $m(t, s) \leq u(t, s, \epsilon)$, $t_0 \leq s \leq s_1 \leq t$. We get for $t_0 \leq s \leq s_1 \leq t$,

$$m(t, s) \leq r(t, s, t_0, u_0) \leq \eta(t, s, s_1, m(t, s_1)).$$

This implies that for $t_0 \leq s \leq s_1 \leq t$,

$$V(s, y(t, s, x(s))) \leq \eta(t, s, s_1, V(s_1, y(t, s_1, x(s_1)))),$$

and thus $x(s) \in \Omega$. Consequently, assumption (4) from Theorem 3.1 yields

$$D_{-}m(t, s_1) \leq g(t, s_1, m(t, s_1)), \tag{3.6}$$

which contradicts (3.5). Therefore, $m(t, s) \leq r(t, s, t_0, u_0)$, $t_0 \leq s \leq t$, since $\lim_{\epsilon \to 0} u(t, s, \epsilon) = r(t, s, t_0, u_0)$ for $t_0 \leq s \leq t < \infty$. It now follows by taking $s = t$ the desired estimate (3.3) and the proof is complete. □

4. Stability Criteria

In this section, we shall discuss sufficient conditions for stability and asymptotic stability of the trivial solutions of (2.1) and (2.2).

THEOREM 4.1. *Assume that*

1. *the hypotheses of Theorem 3.1 hold;*
2. $b(|x|) \leq V(t,x) \leq a(|x|)$, *on* $R_+ \times R^n$, *where* a *and* b *are* K*-class functions* $= [\, \phi \in C[R_+, R_+]\colon \phi(0) = 0$ *and* $\phi(u)$ *strictly increasing in* $u]$;
3. *the trivial solution of (2.1) is uniformly stable.*

Then the uniform stability and uniform asymptotic stability of the trivial solution of (3.2) imply the corresponding uniform stability and uniform asymptotic stability of the trivial solution of (2.2) respectively.

DEFINITION 4.2. When we say the trivial solution of (3.2) is uniformly stable, we mean given $\epsilon > 0$ and $t_0 \in R_+$, there exists a $\delta = \delta(\epsilon) > 0$ such that

$$0 \leq u_0 < \delta$$

implies

$$r_0(t, t_0, u_0) < \epsilon \quad for \quad t \geq t_0,$$

where $r_0(t,t_0,u_0) = r(t,t,t_0,u_0)$ and $r(t,s,t_0,u_0)$ is the maximal solution of (3.2), $0 < s \leq t$. Similarly for uniform asymptotic stability, given $\epsilon > 0$ and $t_0 \in R_+$, there exists a $\delta = \delta(\epsilon) > 0$ and $T = T(\epsilon) > 0$ such that

$$0 \leq u_0 < \delta$$

implies

$$r_0(t, t_0, u_0) < \epsilon \quad for \quad t \geq t_0 + T.$$

PROOF. First, let us assume that the trivial solution of (3.2) is uniformly stable. Using ϵ which is associated with this definition and the function b from assumption (2) from Theorem 4.1, then given $b(\epsilon) > 0$ and $t_0 \in R_+$, there exists a $\delta_1 = \delta_1(\epsilon) > 0$ such that

$$0 \leq u_0 < \delta_1 \quad implies \quad r_0(t, t_0, u_0) < b(\epsilon), \quad t \geq t_0. \tag{4.1}$$

Next, let $\eta = a^{-1}(\delta_1)$. Then assumption (3) from Theorem 4.1 implies given $\eta > 0$, $t_0 \in R_+$, there exists $\delta = \delta(\epsilon) > 0$ such that

$$|x_0| < \delta \quad implies \quad |y(t, t_0, x_0)| < \eta, \quad t \geq t_0. \tag{4.2}$$

We claim that if $|x_0| < \delta$, then $|x(t)| < \epsilon$, $t \geq t_0$. If this is not true, then there would exist a solution $x(t) = x(t, t_0, x_0)$ of (2.2) such that

$$|x_0| < \delta \quad and \quad |x(t_1)| = \epsilon \tag{4.3}$$

for some $t_1 > t_0$ and $|x(t)| < \epsilon$ for $t_0 \leq t < t_1$. By Theorem 3.1, we have the estimate

$$V(t, x(t)) \leq r_0(t, t_0, V(t_0, y(t))), \quad t_0 \leq t \leq t_1, \tag{4.4}$$

which due to (4.1) - (4.4), and assumption (2) from Theorem 4.1 yields

$$\begin{aligned} b(\epsilon) = b(|x(t_1)|) &\le V(t_1, x(t_1)) \le r_0(t_1, t_0, V(t_0, y(t_1))) \\ &\le r_0(t_1, t_0, a(|y(t_1)|)) \\ &\le r_0(t_1, t_0, a(\eta)) = r_0(t_1, t_0, \delta_1) < b(\epsilon)). \end{aligned}$$

This contradiction proves uniform stability of the trivial solution of (2.2).

Now assume that the trivial solution of (3.2) is uniformly asymptotically stable. Then given $b(\epsilon) > 0$ and $t_0 \in R_+$, there exists $\delta_1^* = \delta_1(\rho)$ and $T(\epsilon) > 0$ such that

$$0 \le u < \delta_1^* \ \textit{implies} \ r_0(t, t_0, u_0) < b(\epsilon) \tag{4.5}$$

for $t \ge t_0 + T$. Taking $\epsilon = \rho$, uniform stability of (2.2) yields $\delta_0 = \delta(\rho)$ such that

$$|x_0| < \delta_0 \ \textit{implies} \ |x(t)| < \rho, \quad t \ge t_0. \tag{4.6}$$

To prove uniform asymptotic stability, let $|x(t)| < \delta_0$ and suppose, if possible, that there exists a $\{t_n\}$ with $t_n \ge t_0 + T$ and $t_n \to \infty$ as $n \to \infty$ such that

$$|x_0| < \delta_0 \ \textit{and} \ |x(t_n)| \ge \epsilon \tag{4.7}$$

Again, by Theorem 3.1, we have the estimate

$$V(t, x(t)) \le r_0(t, t_0, V(t_0, y(t))), \quad t \ge t_0. \tag{4.8}$$

As a result, we get using assumption (2) from Theorem 4.1, (4.5), (4.7), and (4.8),

$$\begin{aligned} b(\epsilon) \le b(|x(t_n)|) &\le V(t_n, x(t_n)) \le r_0(t_n, t_0, V(t_0, y(t_n))) \\ &\le r_0(t_n, t_0, a(|y(t_n)|)) \\ &\le r_0(t_n, t_0, \delta_1^*) < b(\epsilon), \end{aligned}$$

which is a contradiction. Therefore, (4.7) is not true which proves that the trivial solution of (2.2) is uniformly asymptotically stable. The proof of the theorem is now complete. □

Now we look at a simple example from [**D**] for illustrative purposes.

EXAMPLE 4.3. Consider the two differential systems

$$y' = e^{-t} y^2, \quad y(t_0) = x_0 \tag{4.9}$$

and

$$x' = e^{-t} x^2 - \int_{t_0}^{t} x \, dx, \quad x(t_0) = x_0 \tag{4.10}$$

Then the solutions of (4.9) are given by $y(t, t_0, x_0) = \frac{x_0}{1 + x_0(e^{-t} - e^{-t_0})}$. There exists a fundamental matrix solution of the corresponding variational equation which is

$$\Phi(t, t_0, x_0) = \frac{1}{[1 + x_0(e^{-t} - e^{-t_0})]^2}.$$

where $\Phi(t, t_0, x_0) = \frac{\partial y}{\partial x_0}(t, t_0, x_0)$. Therefore, if we choose $V(t, x) = x^2$, we get

$$D\_V(t, x) = 2y(t, s, x)\Phi(t, s, x)R(s, x)$$

where $R(t, x) = -\int_{t_0}^{t} x\, dx$ is the perturbation. From (4.10), we see that $g(t, u) = -u^{3/2}$ and hence the solutions of

$$u' = -u^{3/2}, \quad u(t_0) = u_0 \geq 0$$

are $u(t, t_0, u_0) = \frac{4u_0}{[2+u_0^{1/2}(t-t_0)]^2}$. Therefore by Theorem 3.1, we have the relation

$$|x(t, t_0, x_0)|^2 \leq \frac{|x_0|^2}{[1 + x_0(e^{-t} - e^{-t_0} + \frac{t-t_0}{2})]^2}, \quad t \geq t_0$$

which shows that all solutions $x(t, t_0, x_0) \to 0$ as $t \to \infty$, although only some solutions $y(t, t_0, x_0)$ are bounded. For instance, if we set $t_0 = 0$ and $x_0 = 1$, we get that e^t is the corresponding solution of (4.9) where the same initial conditions yields $\frac{2}{2+t+2e^{-t}}$ for (4.10).

There are many exciting things being done in the area of stability analysis. This is just a brief example of some of this work using the variational Lyapunov method.

References

[A] S. G. Deo and V. Lakshmikantham, *Methods of Variation of Parameters for Dynamic Systems*, Gordon and Breach Science Publisher, London 1998.

[B] G. S. Ladde, V. Lakshmikantham, and S. Leela, "A new technique in perturbation theory", Rock Mountain Journal of Mathematics, Vol. 6 (1976), pp. 133 - 140.

[C] V. Lakshmikantham and S. Leela, *Differential and Integral Inequalities*, Vol. 1, Academic Press, New York, 1969.

[D] V. Lakshmikantham, S. Leela, A. A. Martynyuk, *Stability Properties of Nonlinear Systems*, Marcel Dekker, New York 1989.

[E] V. Lakshmikantham, X. Liu, and S. Leela, "Variational Lyapunov method and stability theory", Mathematical Problems in Engineering (to appear).

[F] V. Lakshmikantham and M. Rama Mohana Rao, *Theory of Integro-Differential Equation, Stability and Control: Theory , Methods, and Applications Volume 1.* Gordon and Breach Science Publishers, Switzerland 1995.

Department of Mathematics, Florida A & M University, Tallahassee, FL 32307
E-mail address: `stephens@cis.famu.edu`

Department of Mathematical Sciences, Florida Institute of Technology, Melbourne, FL 32901

II. Papers on Philosophy of Mathematics

Contemporary Mathematics
Volume **275**, 2001

The Art of Mathematics

Jerry P. King

2000 *Mathematics Subject Classification:*

Primary 00A30, Secondary 97D20

Introduction

In two presentations the author described and led discussions on some of the ideas from his book, *The Art of Mathematics.* The author's basic point is that mathematics - like poetry or painting - is art. And by this he does not refer to the use of mathematics in art nor to the use of art in mathematics. M.C. Escher cleverly uses mathematics to produce pictures and patterns of great beauty. Through his sculpture, Helaman Ferguson translates mathematical theorems to produce art objects of marble and bronze. Each of these represents a legitimate connection between mathematics and art. But neither represents the standpoint of *The Art of Mathematics.* In this book, mathematics itself is considered as an art object.

The presentations and the resulting discussions turned on this conception of mathematics as art and on its basic characteristics. There are three characteristics. Mathematics is *an unexpected art, a magical art,* and *a separating art.*

The following is a summary of the basic ideas connected with these notions. Details appear in *The Art of Mathematics (King).*

The Unexpected Art of Mathematics

If mathematics is art then clearly mathematics is an unexpected art. Educated people - if they think of mathematics at all - think of the subject as science. University administrators - who should know better - routinely classify mathematics with science. Thus, for purposes of promotion and salary administration, mathematicians are placed alongside professors from chemistry, physics, and biology. And, since mathematics is not science and has standards vastly different from science, this association works to the detriment of mathematics and of mathematicians.

It is not difficult to see that mathematics is not science. In the first place mathematics is *deductive* while science is *inductive.* The essence of mathematics is its flow from general principles to specific facts. Science works the other way around. Secondly, mathematics consists of theories which have been *proved true.* (Such a theory is called a *theorem.*) On the other hand, a theory is a scientific theory only if it is possible to show it *false.* (I can make a book float in the air when I am alone. I do it by waving my arm. But this theory of motion is not science since it is not *falsifiable.* I can do it only when no one is looking.) Mathematics and science are like the two men in the famous Escher print: they move on a staircase facing the same direction, feet on the same step. But one goes up, the other down. They live in different worlds.

It remains only to show mathematics is art. The quickest way is to apply the *method of appeal to higher authority*, a technique often used by mathematicians in classrooms. And there is no shortage of such authority. For example, former MAA president Lynn Steen says (*Steen*, p.10):

> Aesthetic judgments transcend both logic and applicability in the ranking of mathematical theorems: beauty and elegance have more to do with the value of a mathematical idea than does either strict truth or possible utility.

And Allen L. Hammond says (*Steen*, p.16):

> But mathematicians persist in talking about their field in terms of art - beauty, elegance, simplicity - and draw analogies to painting and music.

Many others - Einstein, Heisenberg, Russell, Hardy, Poincare, etc. - have made the same claim. All of them assert , one way or the other, that mathematics is both motivated and evaluated aesthetically. There is no doubt: mathematics is art. Thus mathematics is an *unexpected art.*

The Magical Art of Mathematics

Like Macbeth's dagger, mathematics lives in the mind. Mathematical objects - things like numbers, equations, matrices - are abstract and imagined. They do not belong to the real world. Certainly, mathematical symbols can be written on pages and printed in books. But the subject itself comes from somewhere deep in the mind. Symbols are merely pictures of ideas. And mathematics is made of ideas. It is not made of ink. The mathematical world of ideas and the real world of physical objects are separate and disjoint. How can there be commerce between them?

None of us doubt the utility of mathematics. From the time of Galileo we have known that the book of nature is written in mathematics. Without the application of mathematics (often the highest and the most abstract mathematics) there would be neither science nor technology beyond simple description and clumsy mechanics. The modern world depends crucially on the application of the most sophisticated mathematics. Yet this application involves the interaction of two different worlds. How can this be? How can the airy nothings of pure mathematics explain and predict natural phenomena?

The applied mathematics process begins with a piece, P, of the real world. (See Figure 1). The problem is to understand P and to learn things about it that are presently unknown. At some stage the examiner- let's say he is a physicist - makes a copy of P over in the mathematical world. The copy is, of course, an *abstraction* or a *model* of P and will consist of certain mathematical objects such as equations, functions, or inequalities. The physicist then manipulates the model using the laws of mathematics and the rules of logic. If he is good enough or lucky enough or both, he learns truths about the model he did not know before. But these new truths are mathematical truths and seemingly apply only to the model. This is the *analysis* step in the process and it involves only *pure mathematics.* This step is not physics. Nor is it any other science.

The physicist now moves from the model back to the real world. He asserts, he claims, or he fervently hopes that the truths he has learned from the model apply to P. This is the *application* step in the process and it has nothing whatever to do with mathematics. Mathematics requires rules of inference. There are no rules which allow the physicist to infer that mathematical truths transfer from the mathematical world to the real world.

Nevertheless, the truths do transfer. More often than not, the mathematical truths carry over to real-world phenomena. Physical objects behave as mathematics says they should. E.P. Wigner (*Wigner*, p.14) says this requires "an article of faith" and he describes the application step of Figure 1 as being *unreasonably effective.*

Yes. But you can say it another way. You might as well call it *magic.* In this step lies the *magical art of mathematics.*

The Separating Art of Mathematics

C.P. Snow gave his famous "Two Cultures" lecture at Cambridge in 1959. In the address Snow argues that all of Western intellectual society has become separated into two groups. He asserts that these two groups consist, respectively, of *the scientists* and *the literary intellectuals.* Snow claims that these groups are separated by "a gulf of mutual incomprehension" and "sometimes hostility and dislike" (*Snow.* p.3).

Snow's lecture touched a nerve and it has been endlessly analyzed. Initially, he was highly, and often bitterly, criticized for daring to suggest that these cultures exist. But his ideas have prevailed and the phrase *the Two Cultures* has become part of the English language.

There is no question of the existence of the two intellectual cultures. You need only visit the humanists on any college campus and ask them their views of science and of scientists. Then go across campus and ask scientists to describe humanists and what they do. You will see Snow's concept of "mutual incomprehension" in action.

However it is possible to produce a more accurate description of the two cultures than Snow's rather vague classification of them as scientists and as literary intellectuals. Chapter 7 of *The Art of Mathematics* argues that the two cultures are more precisely defined by the presence or absence of mathematics. One culture, the *M culture*, consists of those people who possess a certain amount of facility with mathematics. The other, the *N culture*, is composed of those who do not. A precise description of these cultures leans heavily on a turn-of-the-century philosopher named Edward Bullough, his concept of something called *aesthetic distance (King*, p.194), and on the idea of mathematics as art.

Bullough's important idea is this: in the presence of an art object, an observer either experiences the object aesthetically or he does not. Moreover, Bullough claims, if the observer fails to find the object aesthetically pleasing, the failure is one of two types. The failure occurs because the observer is not - in some metaphorical sense - *properly distanced* from the object. He is either too close or else he is too far away.

Bullough thought of these two failures in terms of heat. If the aesthetic distance is too small the object is too hot to be aesthetically pleasing. If the distance is too great the object is too cold to bring pleasure. Examples of under distancing occur frequently with art that involves nudity. Observers often find such objects too hot to be pleasing. Over distancing can occur with art that is highly abstract. The famous squiggle-paintings of Jackson Pollak seem remote to many observers and thus fail to bring pleasure.

It follows from Bullough's argument that each art object has around it - in the sense of aesthetic distance - an *aesthetic ring*. The object provides for an observer an aesthetic experience if and only if the observer lives in the object's aesthetic ring. Otherwise the observer is either too close or too far away and no aesthetic experience exists.

The key point of *The Art of Mathematics* is that mathematics itself may be considered an art object. Thus mathematics possesses an aesthetic ring. Only those inside the ring experience mathematics aesthetically.

Figure 2 shows *the aesthetic ring of mathematics*. This ring, of course, is thin as a dime. No one lives inside except mathematicians. Those with mathematical facility who are not mathematicians are mainly under distanced from mathematics. An industrial engineer, for example, sees mathematics only as a tool for work. The subject reminds him of things undone and projects unfilled. Similarly, a calculus student is too close. Four times each week, mathematics is flung at the student. No way can it bring him pleasure. To both the engineer and the calculus student, mathematics is too hot to be pleasing.

On the other hand, a poet hardly thinks of mathematics at all. The subject is remote from his professional and personal life. Mathematics is as far away as the backside of the moon, as cold as polar ice. The poet is over distanced from mathematics.

These examples extend nicely to the collection of scientists on one hand and to the humanists on the other. Mainly, the scientists are under distanced from mathematics while the literary intellectuals are too far away. And this distancing, this *aesthetic distancing*, precisely defines the Two Cultures of C.P. Snow. You see the cultures in Figure 2. They are separated by the thin aesthetic ring of mathematics. Mathematicians, who live within the ring, form the boundary between the cultures. Thus mathematics defines the barrier between the two cultures. But the barrier could become a bridge.

If you want to eliminate the two cultures it is clear what you must do. Widen the ring. Bring the scientists and the humanists inside.

Teaching

It will not be easy to widen the aesthetic ring of mathematics. Poincare - who was so often correct - is discouraging here. He asserts (see *Papert*, p. 105) that *mathematicians are characterized by the fact that they have the ability to experience mathematics aesthetically.* Moreover, Poincare discourages us further by claiming that *this ability appears only at the highest levels and is innate and thus cannot be learned.* If Poincare is correct then we are stuck with things as they are: the art of mathematics can be appreciated only by mathematicians and the two cultures are forever separate and apart.

Perhaps not. At M.I.T., Seymore Papert recently questioned Poincare's authority (*Papert*, p. 105-119). Papert suggests that Poincare may be correct regarding the significant connection between mathematics and aesthetics but wrong about it being innate and incapable of being learned. Moreover, Papert describes a psychological experiment which partially supports his thesis. That is, Papert has taken a significant first step toward showing two things: *students can learn to see mathematics aesthetically and this learning can take place at elementary levels.*

If this is true then we are speaking of nothing less than a revolution in the teaching of mathematics. Traditionally, mathematics is taught on the basis of its utility. Students are told from first grade to college that they should learn mathematics because it is useful. And, to convince them, they are fretted year after year with dull trivialities disguised as applications. But the students are not convinced. Nor should they be. They know that *to them, and even to most educated people, mathematics is not useful.*

Students do not hear mathematics discussed at the dinner table. Mathematics does not appear on television. Students know of no adults who - beyond simple arithmetic - make any use whatever of mathematics. Consequently, mathematics seems to them - not only the most difficult of subjects - but to carry with it the sour, obnoxious odor of mendacity. As a result, they suffer through the required years of mathematics instruction, all the while making silent vows that when it is done they will never again allow the subject in their presence.

All of us know of the terrible failure of present-day mathematics instruction. The great majority of students despise mathematics and resent being forced to study it. Most of them simply muddle through. And the others, our best students, are only mediocre by world-wide standards.

A teaching revolution is badly needed and one day will come. And the revolution will reject the pedagogy of teaching mathematics simply as a tool for applications because this method has been tried and tried and has been a dismal failure. What form then will the revolution take?

Next time, *let us teach mathematics as art.*

References

1. Jerry P. King, *The Art of Mathematics*, (Plenum , New York 1992)
2. Seymore A. Papert, "The Mathematical Unconscious", in *On Aesthetics and Science,* Judith Weschler ed. (Birkhouser, Boston 1988)
3. C.P. Snow, *The Two Cultures* (Harper and Row, New York 1983)
4. Lynn Steen, ed. *Mathematics Today* (Springer-Verlag, New York 1978)
5. Eugene P. Wigner, "The Unreasonable Effectiveness of Mathematics in the Natural Sciences, *Comm. Pure and Appl. Math. 13 (1980)*

Lehigh University

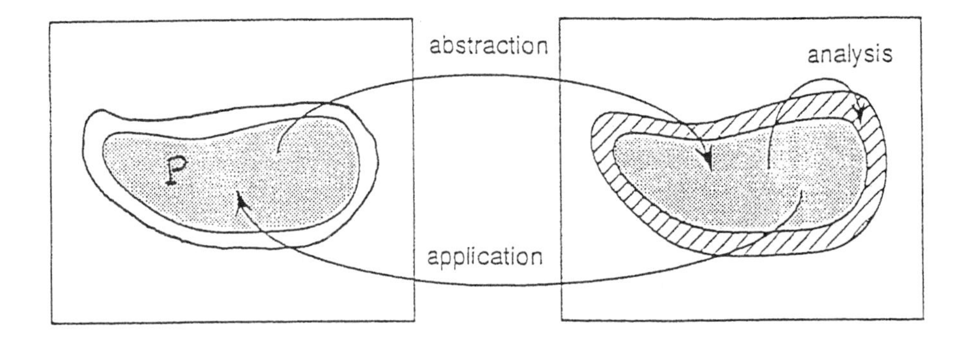

Figure 1: The Applied Mathematics Process

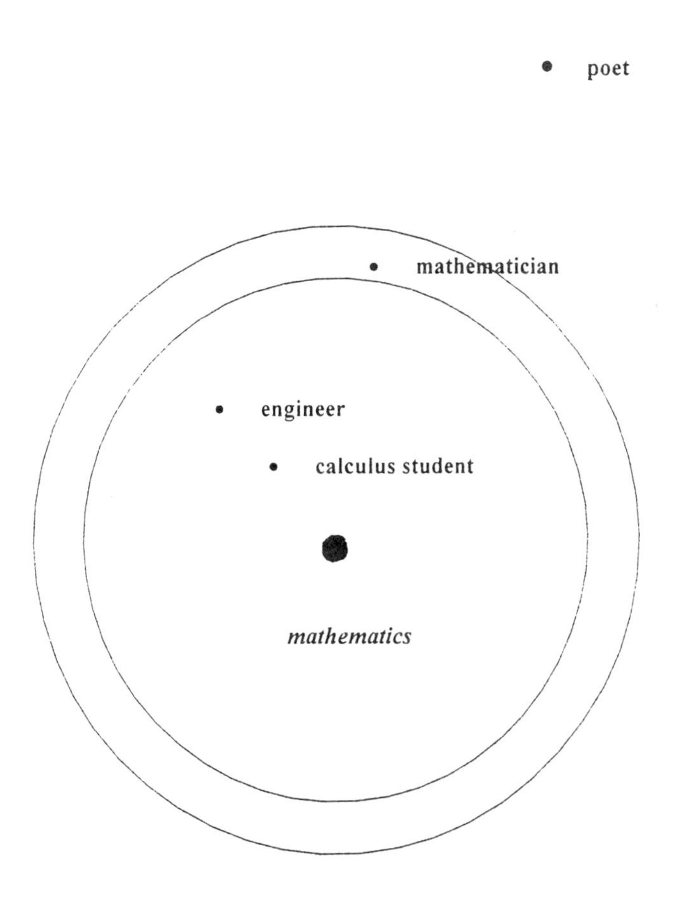

Figure 2: The Aesthetic Ring of Mathematics

Contemporary Mathematics
Volume **275**, 2001

Mathematics is Four Dimensional

William A. Massey

Abstract. To get beyond the archaic 20th century view of pure and applied mathematics, this paper presents a new unifying vision of mathematics as having four dimensions. In the context of this new paradigm, we discuss a proposal for a Professional Masters degree and the development of the Conference for African American Researchers in the Mathematical Sciences.

1. Personal Mathematical Journey

In 1978, I had my first mathematics publication [**7**]. My A. B. degree in mathematics from Princeton University was completed a year earlier, specializing in algebraic number theory. Up-and-coming mathematics and science majors were warned during the 1970's that academic jobs were scarce and it might not be possible to pursue a career in mathematics research. My plan then was to major in "pure mathematics" as an undergraduate. My mathematics classes in college were coupled with physics courses in mechanics, electromagnetism, and quantum mechanics. Pragmatic urges to shift over to "applied mathematics" would be put off until graduate school.

During my senior year of 1977, I applied for a minority fellowship sponsored by Bell Laboratories called the Cooperative Research Fellowship Program[1]. This was a program initiated in 1972 by prominent African-American Bell Labs scientists. Bell Laboratories was one of the few major research institutions that actually had a history of distinguished African-American scientists going back to the 1940's. In hiring Dr. William Lincoln Hawkins in 1942, an African-American chemist and National Engineering medalist winner, Bell Labs broke the "color barrier" five years before major league baseball did so with the hiring of Jackie Robinson in 1947. For the past 28 years, CRFP has helped Bell Laboratories increase the number of African-, Hispanic-, and Native-American Ph.D.'s in the engineering,

2000 *Mathematics Subject Classification.* Primary 00A30, 00B20, 97D20.

Key words and phrases. African American Mathematicians, Professional Masters Program, Conference for African American Researchers in the Mathematical Sciences.

This paper is based on the Albert Turner Bharucha-Reid lecture that I gave for the National Association of Mathematicians at their 1999 Regional Faculty Conference on Research and Teaching Excellence, in Washington D.C. .

mathematical, and physical sciences in the United States (over 120 Ph.D. 's and counting).

In 1977, the year I graduated from college, I was awarded a CRFP fellowship to pursue a Ph.D. in mathematics at Stanford University and work that summer at Bell Laboratories in Murray Hill, New Jersey. One of the successful features of the fellowship was pairing every awardee with a Bell Labs scientist as a mentor. My mentor was James McKenna, head of one of the departments in the Mathematical Sciences Research Center, who introduced me to John Morrison, who in turn introduced me to the applied mathematical field of queueing theory. That summer collaboration led to my first technical paper in 1978.

My experiences at Bell Labs made me eager to learn more about the field of queueing theory. As a graduate student at Stanford University, I took many courses in probability theory, stochastic processes, and applied probability as I was studying for qualifying exams in real analysis, complex analysis and abstract algebra. My Ph.D. dissertation [**6**] was written under the direction of Joseph B. Keller in the area of queueing theory. The main part of the thesis was later published in [**5**]. Since 1981, I have worked at Bell Laboratories and currently have over 50 publications in the fields of queueing theory, performance modelling, telecommunications, and applied probability.

The first thing that my undergraduate training gave me was a sense of the difference between "cookbook calculus" and a rigorous axiom-definition-theorem-proof presentation of calculus, as taught in an honors calculus course by the late Bernard Fox. Reading Michael Spivak's book *Calculus* was an eye-opener and it introduced me to the world of mathematical theory. The second thing I learned was the importance of taking courses in physics. Many mathematical concepts such as cross-products, gradient, curl, divergence, eigenvalues, Hilbert spaces, differential equations, and Fourier transforms, were first encountered in my physics courses. This was especially helpful when taking an honors multivariate calculus course using Michael Spivak's book *Calculus on Manifolds* [**11**] taught by Charles Fefferman. The students with a background in physics, who had seen a "curl" or a "divergence" before could understand the motivation for constructing the exterior derivative of a differential form on a manifold. The ones who had not could do the formal manipulations but had difficultly appreciating why anyone would want to. The third thing I learned was that Princeton considered real analysis, complex analysis, and abstract algebra to be the core material for mathematical theory. This perspective would be reinforced at Stanford since their qualifying exams were structured the same way.

Finally, my introduction to mathematics at Princeton was capped by the experience of writing an expository undergraduate thesis under the direction of the late Bernard Dwork. The thesis was in algebraic number theory and titled "Galois Connections on Local Fields".

My years at Stanford were the beginning of my transition from "pure mathematician" to "applied mathematician". My training through courses and qualifying examination preparation were in pure mathematics and my Ph.D. dissertation topic was in queueing theory, under the direction of an outstanding applied mathematician Joseph Keller. Coupled with spending summers working at Bell Labs, I

[1]The Cooperative Research Fellowship Program was founded in 1972 and is currently co-sponsored by the Lucent Foundation of Lucent Technologies and Bell Laboratories. More information about the program can be found at http://www.bell-labs.com/fellowships/CRFP.

learned that you do not attack an applied mathematics problem by starting with some well known theory and superimpose it on the model. Instead, you start by understanding the model to see what type of mathematics grows organically from it. If the resulting mathematics is standard, then look for the right theorem in the right book. However, if the resulting mathematics in non-standard, then you need to develop the appropriate mathematical theory yourself. Through Joseph Keller, I was fortunate enough to be exposed to a family of queueing models (ones with time-varying rates) that were more realistic as models but lacking in a properly developed mathematical theory. This is the best of all worlds for an applied mathematician, to be able to draw from the world of theoretical mathematics and extend it to model and analyze some new aspect of reality.

My years as a researcher at Bell Labs have given me the opportunity to acquire a deeper understanding about both telecommunications and mathematics (in particular the theory of stochastic processes). This happy synergy came about through understanding the history of queueing theory and comparing the developments between advances in telecommunication systems and advances in queueing models that capture the behavior of these new systems.

Section 2 synthesizes these experiences into the four dimensions of mathematics. The following two sections discuss various applications of this four dimensional theory. Section 3 discusses the development of a professional Masters' program in mathematics and Section 4 discusses the ongoing Conference for African-American Researchers in the Mathematical Sciences (CAARMS).

2. The Four Dimensions of Mathematics

Mathematics needs to lay to rest the "pure" versus "applied" paradigm (purely an artifact of the 20th century) and replace it with a new model. The new approach is one that views the field as a unified, four dimensional whole (see Figure 1). We define the four dimensions of mathematics to be (in alphabetical order): communication, computation, modelling, and theory.

By *communication* we mean the methodology through which mathematical ideas are conveyed to other mathematicians and the world at large. Examples of media that facilitate this type of communication are blackboard and chalk, computer languages, Latex, and pencil and paper. Latex, the mathematical typesetting computer language, is a communication tool used primarily by mathematicians. It is, however, not mathematical theory, computation (at least to the user), or modelling. The products of the communication dimension are books, journal articles, lectures, mathematical notation, and software. It is not surprising that communication is an important dimension, since math has a long distinguished role as the language of science. An underappreciated part of communication is notation. Its importance is easily conveyed through the following example:

EXAMPLE 2.1. Multiply MMMCCVI and LXXIII.

We define *computation* to be the collection of systematic and efficient methodologies for making mathematical calculations. The products of the computation dimension are algorithms with the goal of producing the simplest and fastest algorithms possible.

By *modelling* we mean the fundamental aspects of the real world that are captured and articulated through the language of mathematics. The premier example of the modelling dimension is physics. It has a vast collection of mathematical

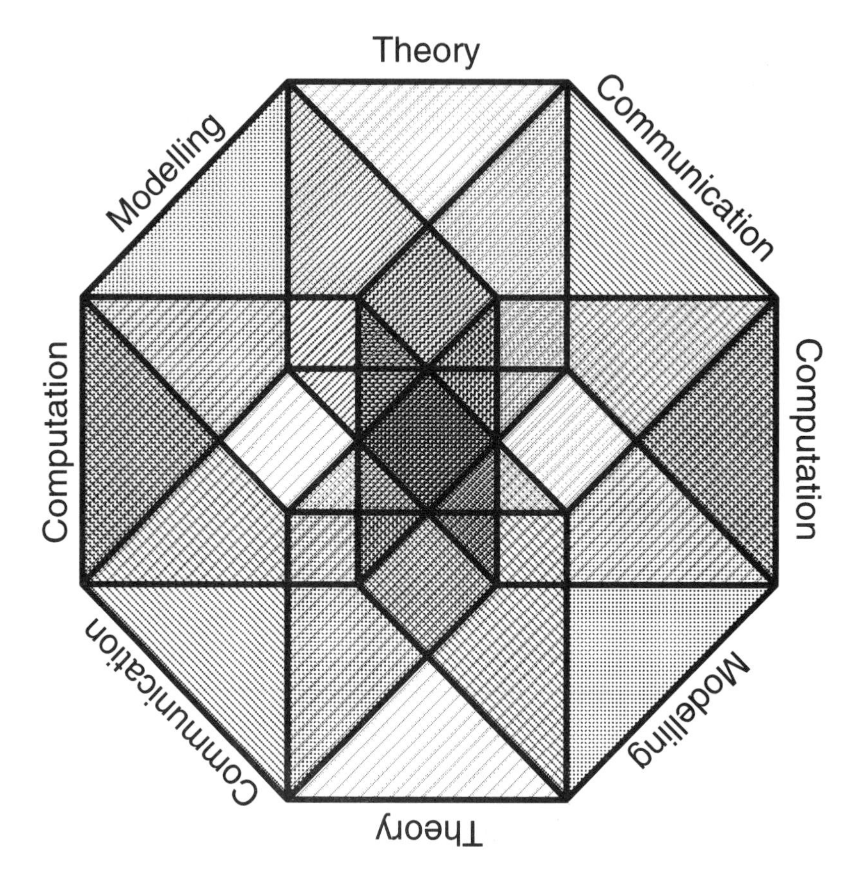

FIGURE 1. The four dimensions of mathematics

tools that were created to describe some physical phenomenon. Einstein's theory of general relativity, for example, can be viewed as the "killer application" for the theory of differential geometry.

Finally, *theory* is well known to be the encoding, description, and summary of mathematical experiences through the narrative of proofs. The theory dimension is one of establishing axioms and producing theorems. One important aspect of theory is developing different types of *intuition*. A given problem may be easier to prove in the context of an algebraic, geometric, physical, or probabilistic intuition. To show the advantages of algebraic intuition, consider the following example:

EXAMPLE 2.2. Picture two planes in Euclidean space that intersect at exactly one point.

This problem is difficult to view geometrically since this is an impossibility within Euclidean three space. At a minimum we must be in Euclidean four space to have this type of intersection. Shifting to an algebraic perspective, construct one of the planes as the set of four-vectors $(x, y, 0, 0)$ for arbitrary scalars x and y. If the second plane is the set of four-vectors $(0, 0, z, w)$ for arbitrary scalars z and w, then it is immediate than these two planes intersect at the single point $(0, 0, 0, 0)$.

Now consider the far reaching theorem below whose proof is immediate when viewed through the lens of geometric intuition.

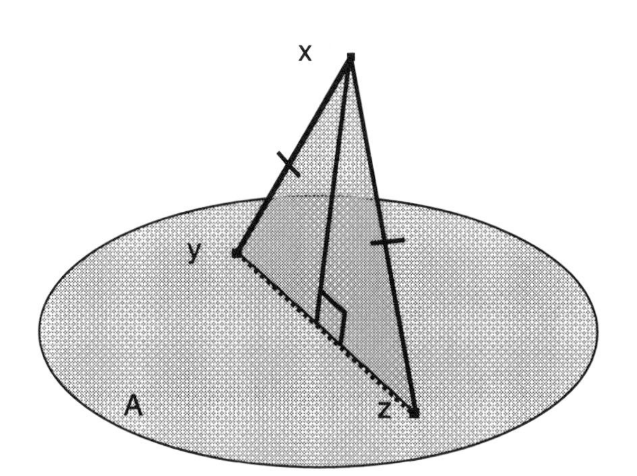

FIGURE 2. Counterproof of uniqueness for minimal distance to convex set

THEOREM 2.3. *Let A be a closed convex subset of a Hilbert space. For all x not belonging to A, there exists a unique y in A that realizes the minimal distance from x to A.*

PROOF. The proof is Figure 2. Assume that y and z are both points in A that minimize the distance from x to any point in A. The points x, y, and z form an isosceles triangle whose two equal sides meet at the vertex x. Since A is a convex set containing both points y and z, then the triangle side opposite the vertex x is a subset of A. Being in a Hilbert space means that we have the notion of orthogonality and so this problem can be solved by high school geometry. Now drop a perpendicular line segment from x to the opposite triangle side. By the triangle inequality, this constructs a line segment from x to a point in A that is shorter that the equal sides of the triangle, if y and z are distinct points. Since this contradicts our premise, we must have $y = z$ and the point that minimizes the distance to A is unique. □

We can extend this multidimensional metaphor by showing that these four dimensions of mathematics are a "linearly independent" set. No three dimensions are a substitute for the remaining one. Discussing the interplay between using software and proving theorems can illustrate these ideas.

Using software is not a substitute for designing software.

The use of software can be thought of as an aspect of mathematical communication. By contrast the act of designing software can be a mixture of theory, modelling, computation, and communication.

Designing software is not an effective substitute for proving theorems.

Creating computational tools to construct a set of computational examples may lead to new mathematical insights and suggest new theorems to prove, but the ultimate theorem must still be proved. Given the computer proof of the Four Color Theorem and the growing computer science field of theorem checking, this view may be challenged in the future. For the present, the case can still be made for needing human beings to prove theorems.

> *Proving theorems is not a substitute for using and designing computational software.*

This is easily verified by asking why a numerical analyst would not solve a set of linear equations numerically by using Cramer's rule.

While these dimensions of mathematics are mutually independent, they also complement each other. We give some examples of what theory contributes to the other dimensions. The following example comes from the world of stochastic processes. We define the random process $\{ N(t) \mid t \geq 0 \}$ to be a *simple point process* if the sample paths of N are increasing step functions with unit jumps where $N(0) = 0$. For a semi-open time interval $(s, t]$ we define the increment of N to be $N(t) - N(s)$. The process N is counting its own jumps so the increment of N for the time interval $(s, t]$ equals the number of jumps that N makes during that interval. The following theorem is due to Prekopa (see [**8**]):

THEOREM 2.4 (Prekopa). *A simple point process is non-homogeneous Poisson if and only if it has independent increments for all mutually disjoint time intervals.*

Even though this theorem was proved in the 1950's, there is still an on-going discussion of the appropriateness of the Poisson process for telecommunication traffic modelling. This theorem clarifies the situation. If we are discussing the arrivals of large number of individual people making calls (or attempted "connections" in the parlance of data traffic), then it is reasonable to make the modelling assumption of independent increments. However, if we are discussing file transfers more microscopically and breaking them down into a transmitted stream of data packets, then there will be correlations between the number of packet arrivals in disjoint time intervals. When a statistical analysis of these arrival streams are made, one actually discovers that the nonhomogeneous Poisson model is not appropriate for modelling the arrival of data packets but it still is an appropriate model for the calling or connection traffic. A recent discussion of these issues can be found in Willinger [**12**].

Below we give an example of how theory informs computation. Consider the algorithm below due to Box, Muller, Marsaglia [**1**] (see also Section 3.4 of Knuth [**4**]). The result follows from the same change-of-coordinates argument that Liouville used (see Spivak [**11**]) to compute the integral of $\exp(-x^2)$ on the entire real line.

THEOREM 2.5 (Box, Muller, Marsaglia). *If we let*

$$r^2 \equiv -2\sigma^2 \log U \quad \text{and} \quad \theta \equiv 2\pi V.$$

where U and V are two independent, uniformly distributed random variables on $[0, 1]$, then X and Y are two independent Gaussian random variables with mean zero and variance σ^2, where

$$X = r\cos\theta \quad \text{and} \quad Y = r\sin\theta.$$

PROOF. Let X and Y be two Gaussian random variables with mean 0 and variance σ^2 and define $\arg(x, y)$ to be the polar coordinate angle between 0 and 2π

for the rectangular coordinate pair (x, y). If we set $s = r^2/2\sigma^2$, then we have

$$\begin{aligned}
&\mathsf{P}\left(X^2+Y^2 > t, 0 \leq \arg(X,Y) < u\right) \\
&= \frac{1}{2\pi\sigma^2}\int\int_{x^2+y^2>t,\,0\leq\,\arg\,(x,y)<u} \exp\left(-\frac{x^2+y^2}{2\sigma^2}\right) dxdy \\
&= \frac{1}{2\pi\sigma^2}\int_0^u\int_{\sqrt{t}}^{\infty} \exp\left(-\frac{r^2}{2\sigma^2}\right) rdrd\theta \\
&= \int_{t/2\sigma^2}^{\infty} \exp(-s)ds \cdot \frac{u}{2\pi} \\
&= \exp(-t/2\sigma^2) \cdot \frac{u}{2\pi}.
\end{aligned}$$

Hence the random variable $X^2 + Y^2$ is exponentially distributed with mean $2\sigma^2$. and it is independent of the random variable $\arg(X, Y)/2\pi$, which is uniformly distributed over the interval $[0, 1]$. □

In the next section we use this four dimensional perspective to discuss the creation of a "professional Masters program" in mathematics.

3. Professional Masters Program

Let us begin this section by asking the following question. There is a bright employment future for mathematics but is there one for mathematicians?

Many young mathematicians express concern about future job opportunities. The recent shortage of academic jobs has led to some strange reactions. The reality is that Ph.D. level mathematics is needed in government and industrial fields such as telecommunications, encryption and national security, as well as finance. Despite this demand, some academicians have made statements such as: "If I cannot secure an academic position for a graduate student who is narrowly trained in my field, then perhaps we should train fewer graduate students". This point of view makes sense only if one accepts the premise that Ph.D. training in mathematics is totally useless for anything except an academic research position in that field. I use this reducto ad absurdum argument to suggest that there is an value in having a deep understanding of mathematics above and beyond the aesthetic values that most mathematicians appreciate.

How to help the future of mathematics? Radical changes in the undergraduate and Ph.D. programs are not necessary. A simple way to improve the training of students in the mathematical sciences is to upgrade the masters program in mathematics. In fields such as engineering or statistics, receiving a masters is an honorable terminal degree. For fields such as mathematics and physics however, the masters degree has the connotation of a "consolation prize" for people who did not obtain a Ph.D. This unfortunate perception needs to be eliminated.

Below is a sketch of a "professional masters program" in mathematics. It is a curriculum that reflects the view of mathematics as an organic four dimensional whole. This program also attempts to provide future math students with skills that will prepare them either for a job or a Ph.D. program in the mathematical sciences. If the Ph.D. path is taken, this may lead to some academic, business, government, or industrial job in the mathematical sciences where having mathematical skills in all four dimensions is critical and will be more so in the future.

The most significant part of the curriculum is the theoretical dimension. We divide it into two core areas. The first core area is analysis:

- Theoretical Calculus (epsilon-delta proofs or non-standard analysis)
- Theoretical Multivariate Calculus (generalized Stokes theorem)
- Real Analysis (measure theory, L^p-spaces)
- Complex Analysis (contour integration, analytic continuation).

All students entering a Masters program will have had courses in calculus, multivariate calculus and differential equations. The first two courses of the analysis core give a consolidated review of this undergraduate material from a rigorous theorem-proof perspective. The two canonical books that cover this type of material are [**10**] and [**11**], both written by M. Spivak. They provide a solid mathematical foundation for learning real and complex analysis. It should be noted that by "real analysis" we mean a course that presents measure theory and the Lebsegue integral.

The second core area is algebra:

- Theoretical Linear Algebra
- Abstract Algebra

This combination allows students to consolidate an understanding of linear algebra from a theoretical perspective and also to see the relationship between this field and abstract algebra. For example, when discussing eigenvalues for matrices, there is a connection between the canonical Jordan normal form for matrices and the decomposition of finitely generated modules over principal ideal domains.

The computational courses will be as follows:

- Computational Programming Language (Examples: C, C++, Fortran)
- Numerical Linear Algebra

A programming course allows mathematicians to see how they can take ideas from the theoretical dimension and translate them into computational algorithms. The course on numerical linear algebra reveals the fundamental role that linear algebra plays in the computational dimension of mathematics. Moreover, this field also reveals how differing goals for the same field affect the ultimate theory that is developed. In the computational realm, returning to the example of matrix eigenvalues, Jordan normal form is viewed as "numerically unstable". However, an insightful theorem, which is rarely found in linear algebra books that emphasize the "algebraic", is found in books on numerical or matrix analysis. This result is the Gershgorin circle theorem, (see Horn and Johnson [**9**]) which we now state:

THEOREM 3.1 (Gershgorin). *Let* $\mathbf{M} = \{ m_{ij} \mid i, j = 1, \ldots, n \}$ *be a matrix with complex entries. Every eigenvalue of* $\mathbf{M}$ *belongs to some circle in the complex plane, centered at some diagonal entry* m_{ii}*, with radius* $\sum_{j \neq i} |m_{ij}|$.

The theorem is a simple inequality, but it communicates significant geometric insight into where the eigenvalues of a matrix are in the complex plane. Moreover, this theorem also communicates insight into the limiting behavior of discrete time Markov chains. Recall that a matrix $\mathbf{P} = \{ p_{ij} \mid i, j = 1, \ldots, n \}$ is defined to be *stochastic* if all its entries are non-negative real and all the row sums equal one. The stochastic matrix $\mathbf{P}$ has one as an eigenvalue which is the spectral radius for all the eigenvalues. Moreover, when every diagonal entry is strictly positive, then

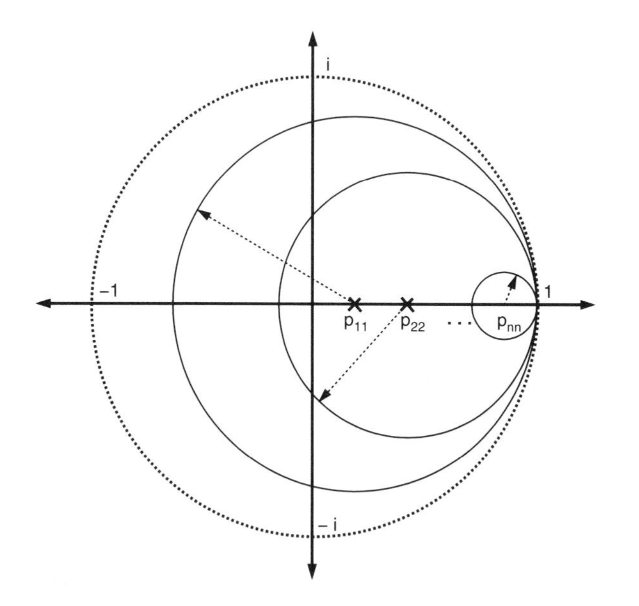

FIGURE 3. Gergorin circles for stochastic matrices

all the other eigenvalues for **P** have a modulus strictly less than one. The proof for this result is geometric (see Figure 3).

The communication topics to be covered are:

- Mathematical Prototyping Language (Examples: Mathematica, Matlab, Maple)
- Mathematical Documentation Language (Example: Latex)
- Webpage Markup Language (Example: HTML)
- Graphical User Interface Language (Examples: Java, Visual Basic, Visual C++)

The first three topics can all be covered in one all-purpose course on "computer numeracy" and the last topics should be a full semester course by itself. The first of these computer languages can be used by mathematicians as a natural extension of pencil and paper for working out examples. The last three computer languages prepare mathematicians for a future where they may be marketing their own work. Computer languages such as Latex allow mathematicians to typeset their own papers and publish their own books. Languages such as HTML allow mathematicians to distribute and promote their papers and books on personal websites that they can create themselves for the internet. Graphical user interface languages such as Java, Visual Basic, or C++ allow mathematicians to transform computational algorithms into "user friendly" software that can be sold to the technical community at large.

Finally, a modelling course for a mathematician simply means taking two semesters of some sufficiently mathematical subject such as

- Physics
- Economics
- Operations Research.

By grounding the mathematical training in a specific context, we see how to abstract phenomena in the "real world" into mathematical principles. In turn we can analyze these principles in the world of mathematics to acquire a deeper insight into the original phenomena.

Now we shape these courses into a two year curriculum. To the course component, we have added an expository Masters thesis and a summer industrial or government internship.

- Year 1 : Semester 1
  - Theoretical Calculus
  - Theoretical Linear Algebra
  - Computational Programming Language
  - Computer Numeracy (Prototyping and Documentation, Webpage Creation)
- Year 1 : Semester 2
  - Theoretical Multivariate Calculus
  - Abstract Algebra
  - Graphical User Interface Language
- Summer Session
  - Industrial/Government Internship or Two Summer Elective Courses
- Year 2 : Semester 1
  - Real Analysis
  - Numerical Linear Algebra
  - Modelling
- Year 2 : Semester 2
  - Complex Analysis
  - Modelling
  - Masters Thesis

In the next section, we discuss the Conference for African American Researchers in the Mathematical Sciences and how it embodies all four dimensions of mathematics.

4. Conference for African American Researchers in the Mathematical Sciences

During my time at Bell Labs, I have met many African-American scientists who are simultaneously prominent researchers in their fields but also have a sense of obligation to encourage more African-American students to pursue Ph.D. 's and excel in the world of science. During the 1980's I noticed that there were research conferences held by and for African-American chemists and similar conferences were held by and for African-American physicists. This motivated me to help create and organize a similar conference in 1995 called the Conference for African American Researchers in the Mathematical Sciences (CAARMS). The conferences have the focus of a small workshop where 12 speakers, drawn predominantly from the African American community, give one-hour talks about their research and 3–4 additional speakers give tutorials. To encourage all the conference participants to attend every talk, no parallel sessions are held The wide range of mathematical subject matter gives the conference the breadth of a large conference. Graduate students are encouraged to contribute technically in a student poster session which has consistently attracted 18–20 contributors. The past 6 conferences from 1995 to 2000 have had roughly 80 people (an even number of researchers and graduate students) in attendance. We have also drawn to the conference over two dozen of the best and brightest young African American researchers. In only 6 years we have had more than 20 African-American graduate students attend a CAARMS conference who have since received a Ph.D. in the mathematical sciences. Some have even returned to CAARMS conferences as Ph.D. researchers to give one of the hour long talks.

Since 1995, I have taken on the role of "perpetual organizer" and have helped to shape the last six CAARMS events. As of this writing, I am working on the seventh CAARMS, which we denote by CAARMS7, with Arlie Petters of Duke University, whose school will also be the hosting site. My past fellow co-organizers of CAARMS conferences have been:

- John R. Birge, Northwestern University (CAARMS5)
- Nathaniel Dean, Rice University (CAARMS2 and CAARMS4)
- Melvin R. Currie, National Security Agency (CAARMS3)
- Arthur D. Grainger, Morgan State University (CAARMS3)
- Donald R. King, Northeastern University (CAARMS6)
- Raymond L. Johnson, University of Maryland (CAARMS1)
- Cassandra M. McZeal, Exxon Mobil Upstream Research Company (CAARMS4)
- Robert E. Megginson, University of Michigan (CAARMS5)
- Richard A. Tapia, Rice University (CAARMS4)
- William P. Thurston, University of California at Berkeley (CAARMS1)
- James C. Turner, Florida State University (CAARMS1)
- Donald C. Williams, Rice University (CAARMS4)
- Pamela J. Williams, Sandia National Laboratories (CAARMS4)
- Leon C. Woodson, Morgan State University (CAARMS3, CAARMS5 and CAARMS6)

Below, we list the past sponsors of the previous six CAARMS conferences:

- Bell Laboratories, Lucent Technologies (CAARMS1, CAARMS2 and CAARMS4)
- Center for Discrete Mathematics and Computer Science (CAARMS2)
- Department of Energy (CAARMS1 and CAARMS4)
- Institute for Advanced Studies (CAARMS2)
- Mathematical Sciences Research Institute (CAARMS1)
- Morgan State University (CAARMS3, CAARMS5 and CAARMS6)
- National Security Agency (CAARMS3, CAARMS4, CAARMS5 and CAARMS6)
- University of Michigan (CAARMS5)
- Sloan Foundation (CAARMS2)

We have published the proceedings for two conferences so far. The first [**2**] was edited by Nathaniel Dean. The second [**3**] was edited by Nathaniel Dean, Cassandra McZeal and Pamela Williams. If you are reading this article, then you are reading a third CAARMS proceedings where Alfred Nöel[2] of the University of Massachusetts at Boston serves as the lead editor.

CAARMS conference attendees get a chance to experience the full four dimensions of mathematics. In a friendly intimate setting, they hear talks about technical areas outside their own expertise only to find mathematical commonalities. The

[2]A fourth CAARMS proceedings is in the works, to be edited by Gaston M. N'guérékata of Morgan State University.

range of talks has gone from gravitational lensing to graph theory, mathematical finance to Lie algebras, the analysis of TCP/IP protocols to algebraic geometry. The juxtaposition of the topics has revealed some surprising similarities and recurring themes among different fields.

We are now formalizing these conference to create an organization called the Council for African American Researchers in the Mathematical Sciences (CAARMS). The CAARMS mission statement is as follows:

Axiom 1: CAARMS exists to encourage, nurture, and promote current and potential African American researchers in the mathematical sciences.

Axiom 2: CAARMS will complement and cooperate with other established groups that promote African-Americans in the mathematical sciences.

Axiom 3: CAARMS will execute these goals by organizing an annual research conference and publishing its proceedings.

References

[1] Box, G. E. P., Muller, M. E., and Marsaglia, G. *Annals Math. Stat.* **28** (1958), 610; and Boeing Scientific Res. Lab. report D1-82-0203 (1962).

[2] Dean, Nathaniel (Editor). *African Americans in Mathematics*, DIMACS: Series in Discrete Mathematics and Theoretical Computer Science, Vol. 34, 1997.

[3] Dean, N., McZeal, C. M., and Williams, P. J. (Editors). *African Americans in Mathematics II*. AMS Contemporary Mathematics Series, Vol. 252., 1999.

[4] *Seminumerical Algorithms: The Art of Computer Programming, Vol. 2.* Addison Wesley Press, 1981, 1969.

[5] Massey, W. A. Asymptotic Analysis of the Time Dependent M/M/1 Queue, *Mathematics of Operations Research*, 10 (May 1985), pp. 305-327.

[6] Massey, W. A. *Nonstationary Queues*, Stanford University, Stanford, California, 1981, PhD Thesis.

[7] Massey, W. A. and Morrison, J. A., Calculation of Steady State Probabilities for Content of Buffer with Correlated Inputs, *Bell System Technical Journal*, 57:9 (November 1978), pp. 3097-3117.

[8] Prékopa, A. On Secondary Processes Generated by a Random Point Distribution of Poisson Type, *Annales Univ. Sci. Budapest de Eötvös Nom. Sectio Math.* **1** (1958) 153–170.

[9] Horn, R. A. and Johnson, C. R. *Matrix Analysis.* Cambridge University Press, 1985.

[10] Spivak, M. *Calculus* (Third Edition). Publish or Perish, Inc., 1994.

[11] Spivak, M. *Calculus on Manifolds.* Addison-Wesley, Publishing Company, 1965.

[12] Willinger, W. and Paxson, V. Where Mathematics Meets the Internet. *Notices of the American Mathematical Society*, Volumne 5, No. 8, September 1998, 961–970. http://www.ams.org/notices/199808/paxson.pdf.

Bell Laboratories, Lucent Technologies, 600 Mountain Avenue, Office 2C–320, Murray Hill, New Jersey 07974

E-mail address: wmassey@lucent.com

III. Tutorials

Contemporary Mathematics
Volume **275**, 2001

COMPACT!
A Tutorial[1]

Scott W. Williams

ABSTRACT. These are the notes from a tutorial on topology presented for the students attending the fifth conference for African Americans in the mathematical sciences. A historical and intuitive approach to highlights of the subject of compact topological spaces are presented.

PREFACE.

I was invited to present a one hour tutorial on Topology for students attending the fifth annual Conference for African American Researchers in the Mathematical Sciences held June 22-26, 1999 at the University of Michigan - Ann Arbor. This five part paper is my notes for the lecture.

When organizer William Massey heard the title "Compact!" of this lecture, he said jokingly, "I hope you will not try to convince us that the finite subcover business is natural." Well, I do not believe it is natural nor do I believe it is intuitive; on the other hand, I believe Topology was invented, in part, to focus upon a few ideas, one of which is extension of the finite to the infinite; another of which is continuity. Compactness was invented to study the former in application to the latter. In this tutorial, I will switch between the intuitive and the accurate, between historical motivations, modern interest and recent results.

What is the importance of compactness? It is used in the definition of the integral; it shows that in linear programming optimal solutions exist on vertices of the feasible set; given a continuous function f from a compact set K to itself, there is an $x \in K$ and an infinite sequence $\langle n_i \rangle$ of integers such that $\lim_{i\to\infty} f^{n_i}(x) = x$. So what is this property "compact?"

1. the beginnings

The Čech Mathematician Bernard Bolzano did a number of remarkable things very early. In particular, in 1817, he extracted numbers from the notion of sequence **[Kline1972]**, and gave an early formulation of finite and infinite sets **[Cantor1883, Jarnik1981]**. In the 1830's he showed that a function continuous on a closed interval is bounded, proved that a bounded sequence has a limit point **[Bolzano1841, Jarnik1981]**, and gave the first example of a continuous nowhere differentiable function (usually attributed to Weierstrass 30 years later). The proof of these essentially led to the Bolzano-Weierstrass Theorem **[Taylor1982]**:

1.1. THEOREM. Every infinite bounded subset of reals has a limit point.

A key to this theorem is an axiom implying that the real line has "no holes except at infinity":

[1] AMS Subject Classification 5402, 54D30 Key Word: Compact

Call the set B. If B contains an infinite increasing sequence, then the least upper bound of the sequence is a limit point. The infinite decreasing sequence case is analogous. By a partitioning B, we see that a bounded set without monotone infinite sequences must be finite.

Before the 19th century, folks already knew that "small" polynomials attained their maxima and minima on closed intervals, but the Bolzano-Weierstrass Theorem led to what turns out to be one of the chief motivations for studying compactness, Weierstrass' theorem **[Taylor 1982]**:

1.2. THEOREM. Each function continuous on a closed subset of a closed interval attains its maximum.

The essence is that the continuous image of a closed and bounded set is closed and bounded. Today we know Theorem1.2 to be true for real-valued functions continuous on any compact space; however, theorems 1.1 and 1.2 suggest an intuitive definition:

1.3. DEFINITION. COMPACT = "NO HOLES": This should be interpreted intuitively. The "proofs" we give in 1.4 are intuitive and need either to be fleshed out with the definition given in 1.5 or the one in 5.2.

1.4. EXERCISES.

1. We recognize two possible kinds of holes - holes at infinity and holes nearby.
a. Loosely, compactness requires a kind of boundedness. ∞ is a hole of the non-negative integers $\omega = \{0,1,2,3,4 \ldots\}$, so ω is not compact. When the hole at ∞ is put in – considering ∞ as a point, the new object $\omega+1$ has no holes.
b. Compactness, loosely, requires a special kind of "closeness." 0 is a hole of the "convergent" sequence $< \frac{1}{2}, \frac{1}{4}, \frac{1}{8}, \frac{1}{16}, \ldots >$ and the sequence, without its limit, is not compact. With its limit, it is compact. More generally, if (0,1] with its usual open sets is declared closed in some topology on the line, then [0,1] is not compact in that topology.
2. Intuitively, "hole" implies some kind of an unending process without resolution. So intuitively, finite sets are compact.
3. Suppose * is a hole of a closed subset F of a compact space X. As a there is stuff of F "close to" *, there is stuff of X "close to" *. But X is compact and * must be a point of X. Since F is a closed subset of X, it contains all points of X close to it. So, a closed subset of a compact set must be compact.
4. A far away hole: The space ω_1 of all countable ordinal numbers has no holes approachable by a countable sequence, yet is not compact. ω_1+1 is the same object with the hole added.

1.5. DEFINITION. 1. A filter is a family Φ of non-empty sets of X which satisfy the condition: If $A,B \in \Phi$, then there is a $C \in \Phi$ contained in $A \cap B$.

2. Here is a correct definition of "close to": A filter Φ converges to (clusters at) the point x provided each neighborhood of x contains (intersects) a member of Φ.

Next is a 1950's definition of compactness closely related to the very first study of compact spaces **[Vietoris1921]**. Its virtues are ease in proving technical results. Its faults are in the shift from points to sets.

3. Here is a correct definition of compact: A space is said to be compact provided each filter consisting of closed sets is contained in a convergent filter; or equivalently, each filter consisting of closed sets clusters at some point.

2. A SPECIAL EXAMPLE

There are some special compact sets: the convergent sequence and its limit, the unit interval and its products, and the unit circle. These tend to dominate how we think of compact objects. However, the Cantor set is a fundamentally important compact object many people believe to be an aberration - it is not. Here's one reason why **[Urysohn 1914]**:

2.1. THEOREM. Each compact metric space is the continuous image of the Cantor set.

THE CANTOR (MIDDLE-THIRDS) **SET** is the set of all real numbers the sum of an infinite series of the form $\sum_{n=1}^{\infty} \frac{2i_n}{3^n}$, where each $i_n = 0$ or 1. Note that if each $i_n = 1$ (=0), then the sum is 1 (0).

Another construction of the Cantor Middle-Thirds Set proceeds recursively via removing various intervals from [0,1]:

Step1: From [0,1]

0 1/3 2/3 1

remove the middle-third open interval $\left(\frac{1}{3}, \frac{2}{3}\right)$ - we get a closed set, picture

C_1: 0 1/3 2/3 1

From both parts of

C_1: 0 1/9 2/9 1/3 2/3 7/9 8/9 1

remove the middle-third open interval - what's left is a closed set, picture

C_2: 0 1/9 2/9 7/9 8/9 1

Again from each of the four parts of

C_2: 0 1/9 2/9 7/9 8/9 1

remove the middle thirds open interval - picture

C_3: 0 ⊢—⊣ ⊢—⊣ ⊢—⊣ ⊢—⊣ ⊢—⊣ ⊢—⊣ ⊢—⊣ ⊢—⊣ 1

And continue .... The Cantor set is the intersection $\mathbf{C} = \cap_{n \in \mathbf{N}} C_n$. As an intersection of closed sets in the compact [0,1] it is closed subset and hence compact. Noticing that the adjacent pairs in **C** can be mapped in an order preserving manner onto the rationals in (0,1). We see the **C** can be also be "pictured" by considering [0,1] and replacing each rational number in (0,1) by two adjacent points. Indeed that picture gives impetus to a special case of Theorem2.1 - [0,1] is the continuous image of **C** {just send adjacent points to one}.

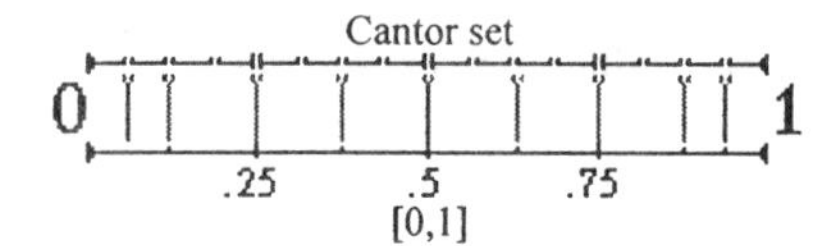

Concerning the Cantor set, one must be careful with intuition. First it is very thin, because the sum of the lengths of the deleted intervals is $\sum_{n=1}^{\infty} \frac{1}{3^n} = 1$; i.e., its measure is zero. On the other hand it has the same size as the entire interval [0,1]. This is strengthened by the problem which appears in W. Rudin's textbook **[W. Rudin 1975]**, and on some Ph.D. Qualifying Exams:

2.2. EXERCISE. Each real in the interval [0,2] is the sum of two members of the Cantor set: For $0 \leq b \leq 2$. Consider the graph L_b of the intersection of line x+y=b with subsets of the square $[0,1]^2$. Indeed, each $C_n{}^2 \cap L_b$ is a non-empty closed set identical to some C_k. Thus, 1.5(3) shows $\mathbf{C}^2 \cap L_b = \cap_{n \in \mathbf{N}} C_n{}^2 \cap L_b \neq \varnothing$. So there are x and y in **C** such that x+y=b.

Recently, the great topologist Mary Ellen Rudin (yes they are married) solved an outstanding problem generalizing 2.1, which asked for a kind of "triangular inequality" extension of metric known as "monotonically normal" **[M. Rudin 1998]**. She proved "sufficient" in:

2.3. THEOREM. In order for a compact space X to be the continuous image of a compact linear ordered space it is necessary and sufficient that for each pair consisting of a point $x \in X$ and its neighborhood G, there exists an open set G_x satisfying two conditions:

1. $x \in G_x \subseteq G$. 2. If $G_x \cap H_y \neq \varnothing$, then either $y \in G$ or $x \in H$.

A pre-print of Rudin's paper is available at the web site TOPOLOGY ATLAS http://at.yorku.ca/topology/ {The "triangular inequality" comment is motivated by observing that in a metric space when B is the open ball about x of radius r, we may take B_x to be the open ball about x of radius $\frac{r}{3}$. Then the triangular inequality proves condition (2)}

3. FUNCTIONAL SEPARATION

Disjoint or non-intersecting closed sets one of which is compact in a (Hausdorff) topological space can be expanded to disjoint open sets - this is called separating them. On the other hand, the distance, in the plane, between the disjoint closed sets, graphs of y=0 and xy=1 is 0. Neither of these sets is compact and ∞ is necessary for this process.

Suppose H and K are disjoint closed sets of a space X. If we can not expand these two to disjoint open sets then there must be some kind of "hole" present. Recall ω and $\omega+1$ from 1.4(1). and ω_1 and ω_1+1 from 1.4(4). Consider the Tychonov Plank, the space X is the product $\omega_1+1\times \omega+1$ with the upper right hand corner removed; i.e., $\omega_1+1\times \omega+1\setminus\{<\infty,\infty>\}$. The closed sets are the top $A = \omega_1\times\{\infty\}$ and the right hand side $B=\{\infty\}\times \omega$.

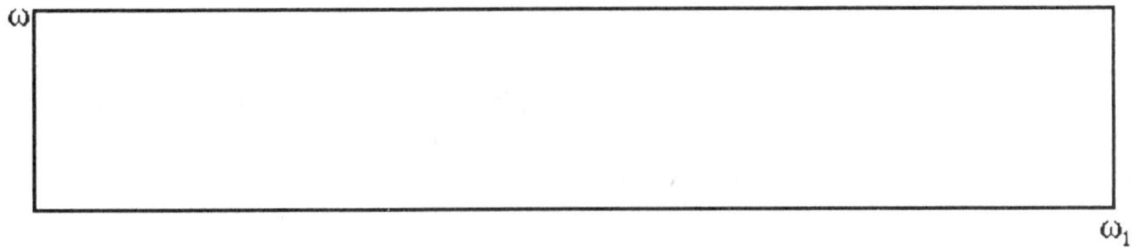

ω

ω_1

{Key to the proof is that the hole at the end of A (see 1.4.5) is so far away that any open set containing B contains countable sequences converging to A; i.e., limit points belonging to A. Thus, there can be no disjoint open sets containing A and B.}

Even stronger (superficially) than "expansion of disjoint closed sets to disjoint open sets" is Urysohn's "separation of closed sets by a continuous function."

3.1. THEOREM. If A and B are disjoint closed subsets of a compact space X there is a continuous $f:X\to[0,1]$ such that f(A)=0 and f(B)=1.

Let us return to products: Fréchet was the first to define a finite product of topological spaces **[Fréchet1910]**. That the product of two compact spaces are compact is intuitively clear: A "hole" in the product of X and Y ought to imply it in a factor. Induction shows "two factors" can be replaced by "finitely many factors." But what about "infinitely many?"

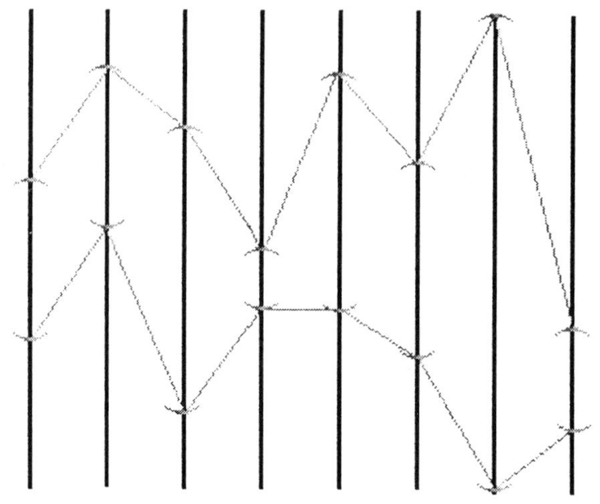

Tietze's product topology

In 1923 Tietze first gave for the general definition of the product of spaces, a topology, now called the box topology, on a product of infinitely many spaces to be that which is generated by the product of open sets **[Tietze1923]** (pictured above). This definition is the "right" way to define product in many areas of mathematics (example, coordinate wise addition in Algebra). However, even the product of countably many two element sets is not compact.

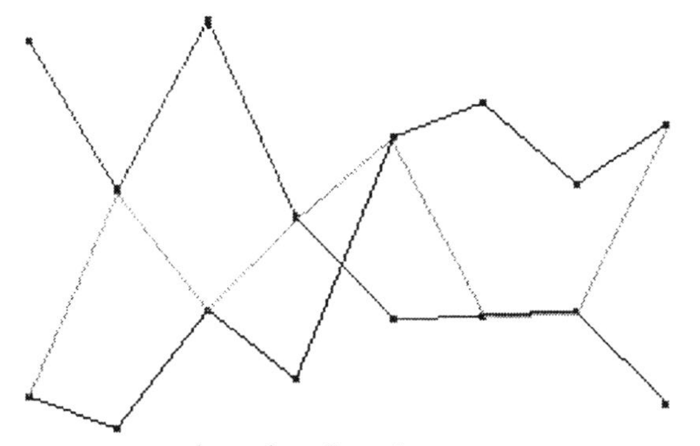

the product of two element sets

Further , an important old and major unsolved problem **[Williams1984]** in topology asks,

Does the product (with the box topology) of countably many copies of [0,1] satisfy the conclusion of Theorem 3.1 ?
The answer to large products is "NO!" in general; i.e., there are disjoint closed subsets A and B of the product $\prod$ of uncountably many copies of [0,1] for which no continuous function f: $\prod \rightarrow [0,1]$ satisfies both f(A)=0 and f(B)=1 **[Lawrence1994]**.

Clearly, Tietze's topology is not good for proving theorems about infinite products (e.g., the preservation of compactness, connectedness, metric etc.), and thus we use a product topology **[Tychonov1930]** which, like compactness, extends finite delicately - the topology is generated by a product of open sets which, only finitely often, may be different from the entire factor:

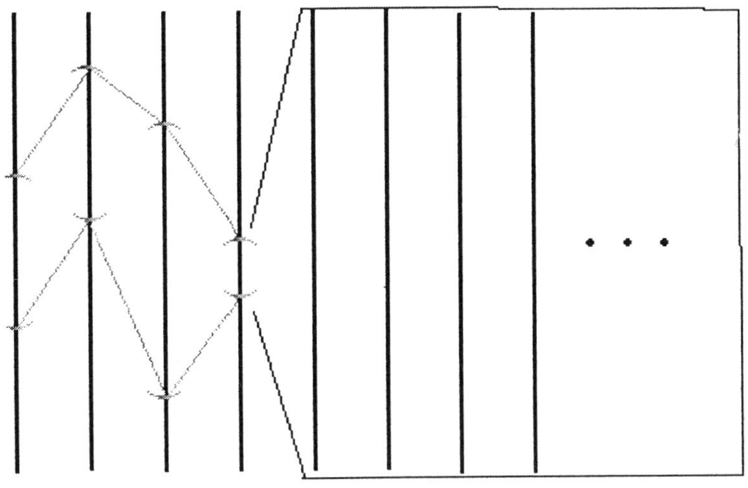

Tychonov product topology

This guarantees that a hole in the product must come from a hole in at least one factor and the standard:

3.2. THEOREM. The product of arbitrarily many compact spaces is compact.

The product of countably many two element spaces is topologically the same as the Cantor set.

4. THE UNIVERSE IN A BOX

Given its derivation from the study of continuity, topology concerns itself with things which are close together while disregarding those which are far apart. Thus, it should be no surprise that we can bound the metric of a space while keeping the topology unaffected by changing the metric to the minimum of 1 and the old distance between two points. In this metric space with a fixed closed set F - the distance d(x,F) between a point x and F forms a continuous function from $X \to [0,1]$ with value 0 on F.

In general, a reasonable axiom for separating points and closed sets in a space X is: given a closed set $F \subseteq X$ and $x \in X \setminus F$ there is a continuous $g:X \to [0,1]$ such that $g(x)=1$ and $g(F)=0$. Using the set **F**, of all continuous $f:X \to [0,1]$, as an index, we build the space P as the product of **F** many copies of [0,1]. The space P is compact and a copy of the space X "sits" in P.

When a "copy" of X sits in P we say X is embedded in P. The above embedding is denoted by ev, for evaluation, and is defined so that the f'th coordinate of ev(x) in the product is f(x).

This tells us when we can consider our space as part of a compact universe:

4.1. THEOREM. In order for a space X to be contained (or embedded) in a compact space it is necessary and sufficient that for each pair consisting of a closed set $F \subseteq X$ and $x \in X \setminus F$ there is a continuous $f: X \to [0,1]$ such that $f(x)=0$ and $f(F)=1$.

4.2. DEFINITION. Suppose a space X is contained (embedded) in a compact space K. Its closure (the set of all points in or close to X) is compact (see 1.3). This closure of a copy of X in a compact space is called a compactification of X. We think of a compactification as filling in the holes of X because we are allowing certain non-convergent filters in X to converge "outside of X."

4.3. EXERCISES. 1. [0,1] is a compactification of (0,1) and hence the reals.
2. The map $t \to \langle \cos 2\pi t, \sin 2\pi t \rangle : (0,1) \to$ unit circle establishes that the unit circle is also a compactification of (0,1) copy of [0,1] with end points identified; i.e., [0,1] and the unit circle, are both compactifications of (0,1) and the reals. Another compactification of (0,1) is the figure 8.
3. There is no continuous function $f: [0,1] \to [-1,1]$ such that $\forall x$, $0<x\leq 1$, $f(x)=\sin(1/x)$; i.e., $\sin(1/x) : (0,1] \to [-1,1]$ has no extension to [0,1]. But by identifying (0,1] with a copy, its graph G in the plane (map each $x \to \langle x,\sin(1/x)\rangle$), we do see that (0,1] is embedded into its compactification $K=G\cup(\{0\}\times[-1,1])$ for which there is a continuous function $f: K \to [-1,1]$ which extends $\sin(1/x)$, namely, project each $\langle x,y\rangle \in K$ to y. K is a compactification of (0,1].
4. Give an example of a compactification K of the integers and continuous function $f: K \to \{0,1,2\}$ which extends $g(n) = n \bmod 3$.

From the view of compactifications, whether one takes the universe to be $\mathbf{R}^3$, $\mathbf{R}^{11}$, $\mathbf{R}^{\infty}$, or one of its subspaces, I say, philosophically, that **the universe is contained in a box** - for example, the compact space P described in the beginning of this section.

4.4. DEFINITION. Prior to Theorem 4.1 above we have nearly described a construction of what is called βX, the Stone-Čech compactification of a non-compact space X. βX is the closure of the "copy" of X and it is called the "largest" compactification **[Willard]** of X because it is characterized by the property that each continuous function f from X to a compact space Y can be extended to a continuous function βf from βX to the same compact space. It is the rule that βX is big - for example, when X is the space of positive integers βX has more points in it than there are reals. For the same reason $\beta(0,1]$ is also quite large. On the other hand, $\beta\omega_1 = \omega_1+1$.

If the universe is contained in a box, an interesting question to consider is "What is left in the box, when we remove the universe?" In other words, what is the nature of βX-X, a research area with I was involved in the early 1980s.

5. OPEN COVERS

Borel proved the following in his 1894 thesis: A countable covering of a closed interval by open intervals has a finite subcover **[Hildebrandt 1924]**. It turns out that Borel's approach was similar to the approach Heine used to prove in 1872 that a continuous function on a closed interval was uniformly continuous (actually first proved, but unpublished for 60 years, by Dirichlet in 1852).
In 1898, Lebesgue (and apparently someone named Cousins in 1895) removed "countable" from the hypothesis of Borel's result. Thus, we have the generalized theorem, which is now commonly called the Heine-Borel theorem, and with modern notation, is:

5.1. THEOREM. A subset of $\mathbf{R}^n$ is compact iff it is closed and bounded.

Unfortunately, this notion of "bounded" does not generalize the theorem for metric spaces, and in topology "metric" need not be present. Vietoris' **[Vietoris1921]** seems to have first seriously considered abstract compact spaces and he proved 1.4(3) and proved "expansion of disjoint closed sets to disjoint open sets" (see the notes before theorem 3.1), but, independently, Alexandrov and Urysohn **[Alexandrov and Urysohn1923]** first gave it the modern definition (though the Russians called the notion "bicompact" for many years):

5.2. DEFINITION. A space X is compact if each covering by open sets contains finitely many open sets which cover.

5.3. EXAMPLES. With the sequence $<\frac{1}{2},\frac{1}{4},\frac{1}{8},\ldots>\cup\{0\}$ or with $\mathbf{N}\cup\{\infty\}$, we see that an open set containing 0 (or ∞}) contains all but finitely many of the points. Thus, an open cover has a finite subcover. To see that [0,1] is compact, use the least upper bound property.

There is an important consequence of compactness which at first appeared to be a property of completeness. It is now known as the Baire Category Theorem due to Baire (1889) for the reals and Hausdorff (1914) for complete metric spaces. It was E. Čech who saw the earlier results were all a consequence of covering properties of certain subsets of a compact spaces **[Čech1937]**:

5.4. THEOREM. Suppose $\{G_n: n\in \mathbf{N}\}$ is a countable family of open dense sets in a compact space. Then the intersection $\bigcap_{n\in \mathbf{N}} G_n$ is dense.

Note that the reals can be embedded as (0,1) into [0,1]. A complete metric space can be embedded as a such an intersection in a compact space. Using 5.4, Banach gave, in 1931, an elegant proof of the Bolzano 1833 **[Jarnik1981]** result usually attributed to Weierstrass (1852):

5.5. COROLLARY. There is a continuous nowhere differentiable function from the reals to the reals.

REFERENCES

Alexandrov, P. and Urysohn, P. (1923) Sur les espaces topologiques compacts, Bull. Intern. Acad. Pol. Sci. Sér. A., 5-8.

Bolzano, B. Functionlehre, 1833-1841 (from Bolzano's manuscripts in Spisy B. Bolzana 1, Prague 1930.

Bolzano, B. Zahlenlehre, 1842 (from Bolzano's manuscripts in Spisy B. Bolzana 2, Prague 1931.

Cantor, G. (1883) Über unendliche lineare Punktmannigfaltigkiet, Math. Annalen 21, 51-58, 545-591.

Čech, E. (1937) On bicompact spaces, Annals of Math. 38, 823-844.

Fréchet, M. (1910) Les dimensions d'un ensemble abstrait, Math. Ann. 68, 145-168.

Hildebrandt, T. H. (1925) The Borel Theorem and its Generalizations. In J. C. Abbott (Ed.), The Chauvenet Papers: A collection of Prize-Winning Expository Papers in Mathematics. Mathematical Association of America.

Jarnik, V. (1981) Bolzano and the foundations of mathematical analysis, Society of Czechoslovak Mathematicians and Physicists.

Kline, M. (1972) Mathematical Thought: From Ancient to Modern Times. Oxford University Press.

Lawrence, L. B. (1996) Failure of normality in the box product of uncountably many real lines. Trans. Amer. Math. Soc. 348, no. 1, 187--203.

Rudin, M.E. (1972) The box product of countably many compact metric spaces, Gen. Top. Appl. 2, 293-248.

Rudin, M. E. (1998) Zero dimensionality and monotone normality, Topology and Applications 85, 319-333.

Rudin, W. (1974) Real and complex analysis. McGraw-Hill Series in Higher Mathematics. McGraw-Hill Book Co., New York.

Taylor, A. (1982). A Study of Maurice Fréchet: I. His Early Work on Point Set Theory and the Theory of Functionals. Archive for History of Exact Sciences, 27 (3), 233-295.

Taylor, A. (1985). A Study of Maurice Fréchet: II. Mainly about his Work on General Topology 1909-1928. Archive for History of Exact Sciences 34 (3), 279-380.

Tietze, H. (1923). Beitrage zur allgemeinen topologie I, Math. Ann. 88, 280-312.

Tychonov, A. (1930). Über die topologische Erweiterung von Räume, Math. Ann. 102, 544-561.

Vietoris, L. (1921). Stetige Mengen, Monatsh. für Math. und Phys. 31, 173-204.

Willard, S. (1968). Generally

Williams, S. (1984), Box products, Handbook of Set-Theoretic Topology (K. Kunen and J.E.Vaughan ed.), North-Holland, 169-200.

Selected Titles in This Series

(Continued from the front of this publication)